OLED DISPLAYS AND LIGHTING

OLED DISPLAYS AND LIGHTING

Mitsuhiro Koden
Yamagata University, Japan

Registered Office
John Wiley & Sons, Ltd, The Atrium, Southern Gate, Chichester, West Sussex, PO19 8SQ, United Kingdom

For details of our global editorial offices, for customer services and for information about how to apply for permission to reuse the copyright material in this book please see our website at www.wiley.com.

Library of Congress Cataloging-in-Publication Data

Names: Koden, Mitsuhiro, author.
Title: OLED displays and lighting / Mitsuhiro Koden.
Description: Chichester, UK ; Hoboken, NJ : John Wiley & Sons, 2016. | Includes bibliographical references and index.
Identifiers: LCCN 2016020346 (print) | LCCN 2016025002 (ebook) | ISBN 9781119040453 (cloth) |
    ISBN 9781119040507 (ePDF) | ISBN 9781119040484 (ePUB) | ISBN 9781119040507 (pdf) |
    ISBN 9781119040484 (epub)
Subjects: LCSH: Light emitting diodes. | Organic semiconductors.
Classification: LCC TK7871.89.L53 K63 2016 (print) | LCC TK7871.89.L53 (ebook) | DDC 621.3815/22–dc23
LC record available at https://lccn.loc.gov/2016020346

A catalogue record for this book is available from the British Library.

Cover image: caracterdesign/gettyimages

Set in 10/12pt Times by SPi Global, Pondicherry, India

Printed and bound in Malaysia by Vivar Printing Sdn Bhd

10 9 8 7 6 5 4 3 2 1

Contents

Preface

Since a bright organic light emitting diode (OLED) device was first reported by C. W. Tang and S. A. VanSlyke of Eastman Kodak in 1987, the high technological potential of OLEDs has been recognized in the display and lighting field. This technological potential of OLEDs has been proved and demonstrated by various scientific inventions, technological improvements, prototypes, and commercial products.

Indeed, OLEDs have various attractive features such as colored or white self-emission, planar and solid devices, fast response speed, thin and light weight, and applicability to flexible applications. Therefore, it should be seen that OLEDs are not only an interesting scientific field but also have great potential for major market applications.

In the past 10 years, OLEDs have experienced serious and complicated business competition from liquid crystal displays (LCDs) and light emitting diodes (LEDs) due to the rapid performance improvement and rapid cost reduction of LCDs and LEDs. However, at the time of writing (2016), new major business possibilities for OLED displays and lighting devices seem to be appearing, in particular, being induced by the huge potential of flexible OLEDs, although LCDs and LEDs are still major devices in displays and lighting, respectively. Rapid growth toward huge market size is forecast for OLED displays and OLED lighting by several market analysts.

For about ten years I have been developing practical OLED technologies at Sharp Corporation, after developing LCD technologies there. Since 2013, I have developed practical flexible OLED technologies at Yamagata University, involving collaborations with a number of private companies.

The purpose of this book is to give an overview of fundamental science and practical technologies of OLEDs, accompanied by a review of the developmental history. This book provides a breadth of knowledge on practical OLED devices, describing materials, devices, processes, driving techniques, and applications. In addition, this book covers flexible technologies, which must be key technologies for future OLED business.

I trust that this book will contribute to not only university students but also researchers and engineers who work in the fields of development and production of OLED devices.

1

History of OLEDs

Summary

Active research and development of OLEDs (organic light emitting diodes) started in 1987, when Tang and VanSlyke of Eastman Kodak showed that a bright luminance was obtained in an OLED device with two thin organic layers sandwiched between anode and cathode. Since their report, OLEDs have been an attractive field from scientific and practical points of view because OLEDs have great potentials in practical applications such as displays and lighting.
This chapter describes the history of the OLED.

Key words

History, Tang, Kodak, Friend, Forrest, Kido, Adachi

Light emission by organic materials was first discovered in a cellulose film doped with acridine orange by Bernanose et al. in 1953 [1]. Ten years later, in 1963, Pope et al., reported that a single organic crystal of anthracene showed light emission induced by carrier injection in a high electric field [2]. Also, since it became known that a large number of organic materials showed high fluorescent quantum efficiency in the visible spectrum, including the blue region, organic materials have been considered as a candidate for practical light emitting devices. However, early studies did not give any indication of the huge potential of OLEDs because of issues such as very high electric field (e.g. some needing 100 V), very low luminance, and very low efficiency. Therefore OLED studies remained as scientific and theoretical fields, not indicating any great motivation towards practical applications.

OLED Displays and Lighting, First Edition. Mitsuhiro Koden.
© 2017 John Wiley & Sons, Ltd. Published 2017 by John Wiley & Sons, Ltd.

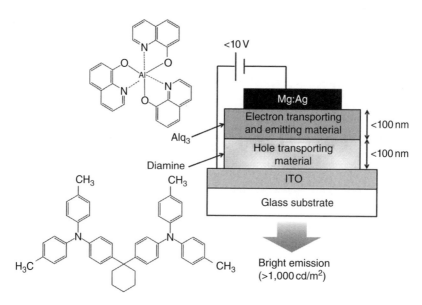

Figure 1.1　The structure and materials of an OLED device reported by Tang et al. [3]

A major impact was made by C. W. Tang and S. A. VanSlyke of Eastman Kodak in 1987. They reported a bright emission obtained in an OLED device with two thin organic layers sandwiched between anode and cathode, as shown in Fig. 1.1 [3]. They introduced two innovative technologies, which used very low thicknesses (<150 nm) of organic layers and adoption of a bi-layer structure. They reported that light emission was observed from as low as about 2.5 V and that high luminance (>1000 cd/m²) was obtained with a dc voltage of less than 10 V. Although the obtained external quantum efficiency (EQE) was still as low as about 1% and the power efficiency was still as little as 1.5 lm/W, the reported results were enough to draw huge attention from scientists and researchers. Indeed, their report started the age of the OLED not only in the academic field but also in industry.

The history of OLEDs is summarized in Table 1.1.

The device reported by Tang and VanSlyke in 1987 consists of a bottom emission structure and small molecular fluorescent monochrome organic material evaporated on glass substrates, but various other novel technologies have been studied and developed, aiming at a revolution in OLED technologies.

In the academic fields, several novel disruptive technologies giving drastic changes in performance of OLEDs have been discovered or invented. These include polymer OLEDs [4], white OLEDs [5], phosphorescent OLEDs [7], multi-photon OLEDs [10], TADF OLEDs [17].

In 1990, Burroughes et al. of the group led by Friend in the Cavendish Laboratories (Cambridge, UK) reported OLED devices with a light emitting polymer [4]. This invention opened the huge possibility of wet-processed OLED technologies.

The first scientific report of white-emission OLED was published in 1994 by Kido et al. of Yamagata University [5]. This report generated active development, aimed at lighting applications for OLEDs. The report also led to developments of the combination of white OLED emission with color filters, aimed at full-color OLED displays.

Table 1.1 The history of OLED

1987	Invention of two-layered OLED with a bright emission (Eastman Kodak / Tang and VanSlyke) [3] (Fig. 1.1)
1990	Invention of polymer OLED (Cavendish Lab. / Burroughes et al. of Friend's group) [4]
1994	First report of white OLED (Yamagata Univ. / Kido et al.) [5]
1997	World's first commercial OLED (Pioneer) [6] (Fig. 1.2) (Passive-matrix monochrome OLED display with bottom emission structure and vacuum deposited small molecular fluorescent materials)
1998	Invention of phosphorescent OLED (Princeton Univ./ Baldo et al. of Thompson and Forrest's group) [7]
1999	Prototype of a full-color polymer AM-OLED display fabricated by ink-jet printing (Seiko-Epson) [8]
2001	Prototype 13″ full-color AM-OLED display (Sony) [9] (Fig. 1.4)
2002	Invention of multi-photon OLED (Yamagata Univ./Kido et al.) [10] Prototype 17″ full-color polymer AM-OLED display fabricated by ink-jet printing (Toshiba) [11]
2003	World's first commercial polymer OLED display (Philips) [12] World's first commercial active-matrix OLED display (SK Display) [13]
2006	Prototype of 3.6″ full-color polymer AM-OLED display with the world's highest resolution (202 ppi) fabricated by ink-jet printing (Sharp) [14]
2007	World's first application of AM-OLED displays for main displays of mobile phones (Samsung) [15] World's first commercial AM-OLED-TV (Sony) [16] (Fig. 1.5)
2009	Invention of TADF (thermally activated delayed fluorescence) (Kyushu Univ. / Endo et al. of Adachi's group) [17]
2010	World's largest OLED display with tiling system using passive-matrix OLED display [18] (Fig. 1.3)
2011	World's first commercial OLED lighting (Lumiotec) [19]
2012	Prototype 13″ flexible OLED display (Semiconductor Energy Laboratory & Sharp) [20] Prototype of 55″ OLED-TV (Samsung) Prototype of 55″ OLED-TV (LG Display)
2013	Prototype of 56″ OLED-TV (Sony) (Fig. 1.6) Prototype of 56″ OLED-TV (Panasonic) World's first commercial flexible OLED lighting with ultra-thin glass (LG Chem) [21] World's first commercial flexible OLED displays (Samsung [22]) World's first commercial flexible OLED displays (LG display [23]) Commercial 55″ OLED-TV (LG Display) [24]
2014	Prototype 77″ OLED-TV with 4 K format (LG Display) [24] Prototype 13.3″ AM-OLED display with 8 K format (Semiconductor Energy Laboratory & Sharp) [25] (Fig. 1.7) Prototype 65″ full-color AM-OLED display fabricated by ink-jet printing (AU Optronics) [26] Commercial flexible OLED lighting using plastic film and roll-to-roll (R2R) production system (Konica Minolta) [27]
2015	Prototype 2.8″ AM-OLED display with ultra high definition format (1058 ppi) (Semiconductor Energy Laboratory) [28] (Fig. 8.15) Prototype 13.3″ foldable AM-OLED display with 8 K format (Semiconductor Energy Laboratory) [29] (Fig. 1.10) Prototype 18″ flexible AM-OLED display (LG Display) [30] (Fig. 10.14) Prototype 81″ Kawara type multi AM-OLED display with 8 K format [31] (Semiconductor Energy Laboratory) (Fig. 10.13)

Figure 1.2 The world's first OLED product (passive-matrix OLED display for car audio) commercialized by Pioneer in 1997 [6]. (provided by Pioneer Corporation)
Display size: 94.7 mm × 21.1 mm
Number of pixels: 16,384 dots (64 × 256)
Color: green (monochrome)
Driving: passive-matrix

In 1998, Baldo et al. of Thompson and Forrest's group at Princeton University reported phosphorescent OLEDs [7], which are theoretically able to realize 100% of internal quantum efficiency. The appearance of phosphorescent OLEDs drastically improved the efficiencies of OLEDs.

In 2002, Kido et al. at Yamagata University reported the multi-photon technology [10], which is able to realize high luminance and long lifetime.

In 2011, Endo et al. of the group led by Adachi in Kyushu University reported thermally activated delayed fluorescence (TADF) [17], which is an alternative technology to phosphorescent OLEDs for realizing high efficiency.

In parallel with such inventions and discoveries, much effort has been devoted to technological development on topics such as performance improvement, analysis of emission mechanism and degradation.

Much effort has also been devoted to practical device development and commercialization.

In 1997, Pioneer commercialized the world's first OLED display, which was a passive-matrix green monochrome display (Fig. 1.2) [6]. This display had a bottom emitting monochrome device structure fabricated using vacuum evaporation technology with small molecular fluorescent organic materials. This display was applied to car audio.

The world's first polymer type OLED was commercialized by Philips in 2003 [12]. This was a passive-matrix yellow monochrome display, being applied in shavers.

Currently, passive-matrix OLED displays are widely used in various applications with small to medium information content. In addition, tiling technology with passive-matrix OLED displays has realized very large size display (e.g. 155″) and cubic type displays such as a terrestrial globe display as shown in Fig. 1.3 [18].

Active-matrix OLED (AM-OLED) displays with full-color images have also been actively developed. In 2001, Sony demonstrated a 13″ active-matrix full-color OLED display with 800 × 600 pixels (SVGA), which had a major impact on the display industry (Fig. 1.4) [9]. The OLED display was constructed using several novel technologies: top-emission structure with micro-cavity design for increasing luminance and they achieved excellent color purity, novel

Figure 1.3   The world largest cubic-type terrestrial globe display "Geo-Cosmos" [18]. (provided by Miraikan, the National Museum of Emerging Science and Innovation, Japan)

Figure 1.4   A 13″ active-matrix full-color OLED display developed by Sony in 2001 [9]. (provided by Sony Corporation)
Display size: 13″ (264 mm × 198 mm)
Number of pixels: 800 × 600 (SVGA)
Color: full color, R(0.66,0.34)   G(0.26,0.65)   B(0.16,0.06)
Luminance: higher than 300 cd/m^2
Driving: Active-matrix with LTPS

current-drive LTPS-TFT circuit with four TFTs for attaining uniform luminance over the entire screen, solid encapsulation for enabling thinner structure, etc. In addition, the display was the largest OLED display at that time, and the pictures with beautiful color – R(0.66,0.34), G(0.26,0.65), B(0.16,0.06) – high luminance (>300 cd/m²), high contrast ratio, and wide viewing angle, greatly impressed many scientists and researchers in OLED and display fields.

The world's first active matrix OLED display was commercialized in 2003 by SK Display (a joint company by Eastman Kodak and Sanyo Electric) [13]. The display was used by Kodak in digital cameras.

Sony commercialized the world's first OLED-TV in 2007 (Fig. 1.5) [16]. The display size was 11″ diagonal. In mobile application, in 2007, Sumsung's full-color AM-OLED displays were applied to main displays of mobile phones [15].

In 2012 and 2013, 55 or 56 inch large size OLED-TVs were demonstrated by Samsung, LG Display, Sony (Fig. 1.6), Panasonic, respectively. LG Display commercialized a 55″ OLED-TV [24] in 2013 and developed a 77″ OLED-TV prototype in 2014 [24]. In addition, Semiconductor Energy Laboratory (SEL) developed high resolution AM-OLED displays, such as a 13.3″ AM-OLED display with 8 K format (664 ppi) (Fig. 1.7) in 2014 [25] and a 2.8″ AM-OLED display with ultra high definition format (1058 ppi) (see Fig. 8.15) in 2015 [28].

On the other hand, AM-OLEDs using light emitting polymers have also been developed. In 1999, Seiko Epson demonstrated an ink-jet AM-OLED display [8], in 2002, Toshiba showed a 17″ prototype polymer AM-OLED fabricated by ink-jet [11], and in 2006, Sharp demonstrated a polymer AM-OLED with the world's highest resolution (202 ppi), fabricated by ink-jet printing [14]. Moreover, in 2014, AU Optronics presented a 65″ AM-OLED prototype display fabricated by ink-jet printing [26].

Figure 1.5 The world's first OLED-TV commercialize by Sony in 2007 [16]. (provided by Sony Corporation)
Display size: 11″ (251 mm × 141 mm)
Number of pixels: 960 × 540 (QHD)
Color: full color
Contrast ratio: higher than 1,000,000 : 1
Driving: active-matrix with LTPS

Figure 1.6 A 56″ active-matrix full-color OLED display with 2 K4 K format developed by Sony in 2013. (provided by Sony Corporation)

Figure 1.7 Prototype of 13.3″ AM-OLED display with 8 K format (Semiconductor Energy Laboratory & Sharp) [25]. (provided by Semiconductor Energy Laboratory)
Display size: 13.3 inches (165 mm × 294 mm)
Number of pixels: 4320 × 7680 (4 k8 k)
Resolution: 664 ppi
Color: full color
Device structure: white tandem OLED (top emission) + color filter
Driving: active-matrix with CAAC-IGZO TFT

Figure 1.8 Examples of OLED lighting. (provided by Lumiotec Inc.)

Figure 1.9 Examples of OLED lighting. (provided by Kaneka Corporation)

In the lighting fields, in 2011, Lumiotec, which is a Japanese venture company for OLED lighting, commercialized the world first OLED lighting [19]. At present, active demonstrations have shown new usage images of OLED lighting by several companies, as shown in Figs 1.8 and 1.9.

Recently, flexible OLEDs have also been actively developed. In 2011, Semiconductor Energy Laboratory (SEL) developed a 3.4″ flexible OLED prototype display with 326 ppi [32]. SEL and Sharp developed a 13.5″ flexible OLED prototype display with 81.4 ppi in 2012 [20] and SEL developed a 13.3″ flexible OLED prototype display with 8 K format (664 ppi) in 2015 [29]. Figure 1.10 shows a prototype of 13.3″ foldable AM-OLED display with 8 K format. The world's first flexible OLED displays were commercialized by Samsung [22] and LG Display [23]. In addition, in 2015, LG Display developed an 18″ prototype flexible display [30] and SEL developed an 81″ Kawara type multi-AM-OLED display with 8 K format (see. Fig. 10.13) [31].

On the other hand, world's first flexible OLED lighting was commercialized by LG Chem in 2013, using flexible ultra-thin glass [21]. In 2014, Konica Minolta commercialized flexible

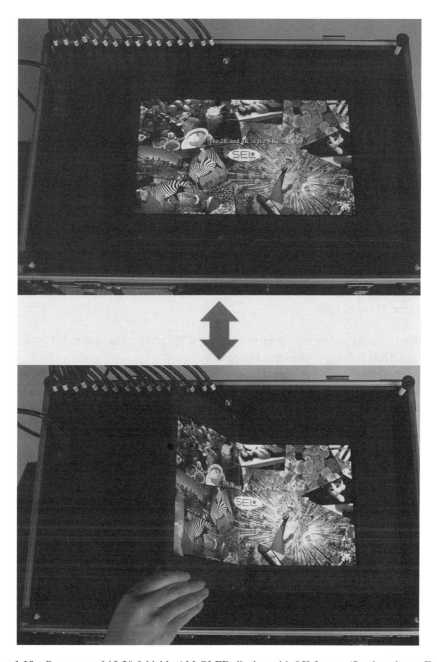

Figure 1.10   Prototype of 13.3″ foldable AM-OLED display with 8 K format (Semiconductor Energy Laboratory) [29]. (provided by Semiconductor Energy Laboratory)
Display size: 13.3″ (165 mm × 294 mm)
Number of pixels: 4320 × 7680 (4 k 8 k)
Resolution: 664 ppi
Color: full color
Device structure: white tandem OLED (top emission) + color filter
Driving: active-matrix with CAAC-IGZO TFT

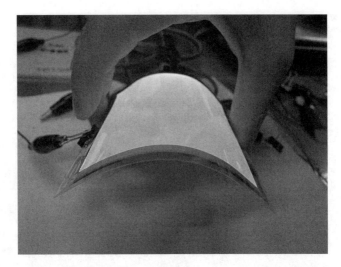

Figure 1.11 A prototype flexible OLED lighting panel [33]. (provided by NEC Lighting, Ltd)
Panel size: 92 mm × 92 mm
Emission area: 75 mm × 75 mm

OLED lighting using plastic film and a roll-to-roll (R2R) production system [27]. An example of flexible OLED lighting is shown in Fig. 1.11.

As described above, various OLED technologies have been actively developed since 1987 and OLED products were steadily commercialized in competition with LCDs (liquid crystal displays) or LEDs (light emitting diodes). While the current business situation of OLEDs is not so favorable due to their high cost, etc., OLEDs have huge potential business in combination with further effort on cost reduction. In addition, "flexible" is a very significant key word for OLEDs because flexible OLEDs can realize certain commercial products that LCDs and LEDs cannot. Current active research and development of flat and flexible OLEDs should be understood in the light of this huge potential for future business.

References

[1] A. Bernanose, M. Comte and P. Vouaux, *J. Chem. Phys.*, **50**, 64–68 (1953).
[2] M. Pope, H. P. Kallmann and P. Magnante, J. Chem. *Soc.*, **38**, 2042–2043 (1963).
[3] C. W. Tang and S. A. VanSlyke, *Appl. Phys. Lett.*, **51**, 913–915 (1987).
[4] J. H. Burroughes, D. D. Bradley, A. R. Brown, R. N. Markes, R. Mackay, R. H. Friend, P. L. Burns and A. B. Holmes, *Nature*, **347**, 539–541 (1990).
[5] J. Kido, K. Hongawa, K. Okuyama, K. Nagai, *Appl. Phys. Lett.* **64**, 815–817 (1994).
[6] T. Wakimoto, R. Murayama, K. Nagayama, Y. Okuda, H. Nakada, T. Tohma, *SID 96 Digest*, 849 (1996).
[7] M. A. Baldo, D. F. O'Brien, Y. You, A. Shoustikov, S. Sibley, M. E. Thompson, and S. R. Forrest, *Nature*, **395**, 151–154 (1998).
[8] T. Shimoda, M. Kimura, S. Miyashita, R. H. Friend, J. H. Burroughes, C. R. Towns, *SID 99 Digest*, 26.1 (p. 372) (1999); T. Shimoda, S. Kanbe, H. Kobayashi, S. Seki, H. Kiguchi, I. Yudasaka, M. Kimura, S. Miyashita, R. H. Friend, J. H. Burroughes, C. R. Towns, SID 99 Digest, 26.3 (p. 376) (1999).
[9] T. Sasaoka, M. Sekiya, A. Yumoto, J. Yamada, T. Hirano, Y. Iwase, T. Yamada, T. Ishibashi, T. Mori, M. Asano, S. Tamura, T. Urabe, *SID 01 Digest*, 24.4 L (p. 384) (2001); J. Yamada, T. Hirano, Y. Iwase, T. Sasaoka, *Proc. AM-LED'02*, OD-2 (p. 77) (2002); www.sony.net/SonyInfo/News/Press_Archive/200102/01-007aE/

[10] J. Kido, J. Endo, T. Nakada, K. Mori, A. Yokoi and T. Matsumoto, *Ext. Abstract of "Japan Society of Applied Physics, 49th Spring Meeting"*, 27p-YL-3 (p. 1308) (2002); T. Matsumoto, T. Nakada, J. Endo, K. Mori, N. Kawamura, A. Yokoi, J. Kido, *SID 03 Digest*, 979 (2003).

[11] M. Kobayashi, J. Hanari, M. Shibusawa, K. Sunohara, N. Ibaraki, *Proc. IDW'02*, AMD3-1 (p. 231) (2002).

[12] M. Fleuster, M. Klein, P. v. Roosmalen, A. d. Wit, H. Schwab, *SID 04 Digest*, 44.2 (p. 1276) (2004).

[13] K. Mameno, R. Nishikawa, K. Suzuki, S. Matsumoto, T. Yamaguchi, K. Yoneda, Y. Hamada, H. Kanno, Y. Nishio, H. Matsuola, Y. Saito, S. Oima, N. Mori, G. Rajeswaran, S. Mizukoshi, T. K. Hatwar, *Proc. IDW'02*, 235 (2002).

[14] T. Gohda, Y. Kobayashi, K. Okano, S. Inoue, K. Okamoto, S. Hashimoto, E. Yamamoto, H. Morita, S. Mitsui, M. Koden, *SID 06 Digest*, 58.3 (p. 1767) (2006).

[15] News release of KDDI, 20 March 2007. www.kddi.com/corporate/news_release/2007/0320/

[16] News release of Sony Corporation, 1 October 2007. www.sony.jp/CorporateCruise/Press/200710/07-1001/

[17] A. Endo, M. Ogasawara, T. Takahashi, D. Yokoyama, Y. Kato, C. Adachi, *Adv. Mater.*, **21**, 4802 (2009); A. Endo, K. Sato, K. Yoshimura, T. Kai, A. Kawada, H. Miyazaki, and C. Adachi, *Appl. Phys. Let.*, **98**, 083302 (2011).

[18] Z. Hara, K. Maeshima, N. Terazaki, S. Kiridoshi, T. Kurata, T. Okumura, Y. Suehiro, T. Yuki, *SID 10 Digest*, 25.3 (p. 357) (2010); S. Kiridoshi, Z. Hara, M. Moribe, T. Ochiai, T. Okumura, Mitsubishi Electric Corporation Advance Magazine, **45**, 357 (2012).

[19] News release of Lumiotec, 24 July 2011. www.lumiotec.com/pdf/110727_LumiotecNewsRelease%20JPN.pdf

[20] S. Yamazaki, J. Koyama, Y. Yamamoto, K. Okamoto, *SID 2012 Digest*, 15.1 (p. 183) (2012).

[21] News release of LG Chem, 3 April 2013. www.lgchem.com/global/lg-chem-company/information-center/press-release/news-detail-527

[22] News release of Samsung Electronics, 9 October 2013. http://global.samsungtomorrow.com/?p=28863

[23] S. Hong, C. Jeon, S. Song, J. Kim, J. Lee, D. Kim, S. Jeong, H. Nam, J. Lee, W. Yang, S. Park, Y. Tak, J. Ryu, C. Kim, B. Ahn, S. Yeo, *SID 2014 Digest*, 25.4 (p. 334) (2014).

[24] C.-W. Han, J.-S. Park, Y.-H. Shin, M.-J. Lim, B.-C. Kim, Y.-H. Tak, B.-C. Ahn, *SID 2014 Digest*, 53.2 (p. 770) (2014).

[25] S. Yamazaki, *SID 2014 Digest*, 3.3 (p. 9) (2014);S. Kawashima, S. Inoue, M. Shiokawa, A. Suzuki, S. Eguchi, Y. Hirakata, J. Koyama, S. Yamazaki, T. Sato, T. Shigenobu, Y. Ohta, S. Mitsui, N. Ueda, T. Matsuo, *SID 2014 Digest*, 44.1 (p. 627) (2014).

[26] P.-Y. Chen, C.-L. Chen, C.-C. Chen, L. Tsai, H.-C. Ting, L.-F. Lin, C.-C. Chen, C.-Y. Chen, L.-H. Chang, T.-H. Shih, Y.-H. Chen, J.-C. Huang, M.-Y. Lai, C.-M. Hsu, Y. Lin, *SID 2014 Digest*, 30.1 (p. 396) (2014).

[27] T. Tsujimura, J. Fukawa, K. Endoh, Y. Suzuki, K. Hirabayashi, T. Mori, *SID 2014 Digest*, 10.1 (p. 104) (2014).

[28] K. Yokoyama, S. Hirasa, N. Miyairi, Y. Jimbo, K. Toyotaka, M. Kaneyasu, Y. Miyake, Y. Hirakata, S. Yamazaki, M. Nakada, T. Sato, N. Goto, *SID 2015 Digest*, 70.4 (p. 1039) (2015).

[29] K. Takahashi, T. Sato, R. Yamamoto, H. Shishido, T. Isa, S. Eguchi, H. Miyake, Y. Hirakata, S. Yamazaki, R. Sato, H. Matsumoto, N. Yazaki, *SID 2015 Digest*, 18.4 (p. 250) (2015).

[30] J. Yoon, H. Kwon, M. Lee, Y.-y.l Yu, N. Cheong, S. Min, J. Choi, H. Im, K. Lee, J. Jo, H. Kim, H. Choi,Y. Lee, C. Yoo, S. Kuk, M. Cho, S. Kwon, W. Park, S. Yoon, I. Kang, S. Yeo, *SID 2015 Digest*, 65.1 (p. 962) (2015).

[31] D. Nakamura, H. Ikeda, N. Sugisawa, Y. Yanagisawa, S. Eguchi, S. Kawashima, M. Shiokawa, H. Miyake, Y. Hirakata, S. Yamazaki, S. Idojiri, A. Ishii, M. Yokoyama, *SID 2015 Digest*, 70.2 (p. 1031) (2015).

[32] K. Hatano, A. Chida, T. Okano, N. Sugisawa, T. Nagata, T. Inoue, S. Seo, K. Suzuki, M. Aizawa, S. Yoshitomi, M. Hayakawa, H, Miyake, J. Koyama, S. Yamazaki, Y. Monma, S. Obana, S. Eguchi, H. Adachi, M. Katayama, K. Okazaki, M. Sakakura, *SID 2011 Digest*, 36.4 (p. 498) (2011).

[33] M. Koden, H. Kobayashi, T. Moriya, N. Kawamura, T. Furukawa, H. Nakada, *Proc. IDW'14*, FMC6/FLX6-1 (p. 1454) (2014).

2

Fundamentals of OLEDs

Summary

This chapter describes fundamentals of OLEDs, including the principles, the fundamental device structures and some features of OLEDs.

OLED is a carrier injection type electroluminescent device, having a similar emission mechanism to the inorganic light emitting diode (LED). The fundamental device structure of OLEDs is very simple, consisting of an anode, some organic layers and a cathode. The OLED has several attractive features such as low driving voltage, high efficiency, wide color variation, and fast response speed.

Key words

principle, device, structure, feature

2.1 Principle of the OLED

In OLED devices, organic layers are sandwiched between two electrodes, an anode and a cathode as shown in Fig. 2.1. The organic layers usually consist of multiple layers, in which each layer plays an intrinsic role. When a voltage is applied to an OLED device, charge carriers are injected from the electrodes to the organic layers. A hole (positive charge) is injected from the anode and an electron (negative charge) is injected from the cathode. The holes and electrons are transported to an emission site and recombined. Organic materials in the emission site are excited by recombination of holes and electrons. And then emission occurs when the excited state goes back to the ground state.

Such mechanism can also be illustrated by using an energy diagram as shown in Fig. 2.2.

OLED Displays and Lighting, First Edition. Mitsuhiro Koden.
© 2017 John Wiley & Sons, Ltd. Published 2017 by John Wiley & Sons, Ltd.

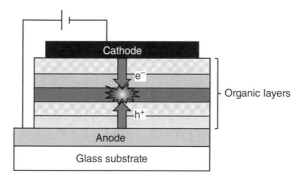

Figure 2.1   Schematic illustration of a typical OLED device structure and the emission mechanism

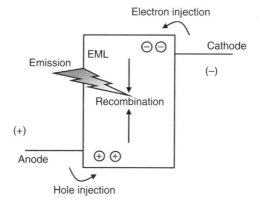

Figure 2.2   OLED emission mechanism illustrated by using an energy diagram

Emission itself is simply illustrated in Fig. 2.3, which shows energy absorption and energy release between two states with different energy levels (*Em* and *En*). The emission is represented by the following equation, where *h* is the Plank constant and ν is the frequency of the emission.

$$hv = |Em - En| = \Delta E$$

According to this equation, the emission frequency in OLEDs is dependent upon the energy gap between the excited and ground states. Practically speaking, emission colors can be controlled by the energy gap between the excited and ground states.

As described above, the OLED is a carrier injection type electroluminescent device using organic materials. Therefore, the OLED is sometimes called "organic electroluminescent", "organic EL", "OEL", etc. However, it should be noticed that the emission mechanism of the OLED is essentially different from an inorganic electroluminescent device which is usually called "EL". On the other hand, the emission mechanism of the OLED is similar to inorganic light emitting diode, which is usually called an "LED". From this point of view, it can be said that "organic light emitting diode (OLED)" is a suitable name.

There are three common types of emission mechanisms in OLEDs. They are fluorescent OLED, phosphorescent OLED, and the thermally activated delayed fluorescent (TADF) OLED. In Chapter 3, these light emission mechanisms are described in more detail.

2.2   Fundamental Structure of the OLED

An OLED is a type of solid device, being different from CRT (cathode ray tube or "Braun tube"), PDP (plasma display panel), LCD (liquid crystal display), etc. A typical device structure of OLED is shown in Fig. 2.4, although various types of OLED devices are possible. Organic layers are sandwiched between an anode and a cathode. Holes and electrons are injected from anode and cathode, respectively. As already described, organic materials are excited by recombination of holes and electrons and then emission occurs when the excited state returns to the ground state.

In order to realize practically valuable performances such as high efficiency, long lifetime, and required colors, a number of successive layers with different function are usually stacked. In such multi-layer structures, the various layers play such roles such as charge injection, charge transport, emission, and charge blocking. Therefore, these layers are often called hole injection layer (HIL), hole transport layer (HTL), emission layer (EML), electron transport layer (ETL), and electron injection layer (EIL), as shown in Fig. 2.4.

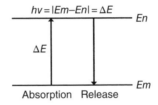

Figure 2.3   Schematic illustration of emission from the excited state

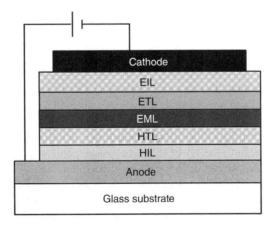

Figure 2.4   A typical device structure of OLEDs

The anode is required to have a high work function for hole injection. Currently, indium tin oxide (ITO) is a commonly used anode material. The work function of ITO is around 4.7 eV, which is not high enough to inject a hole into organic materials because HOMO levels of the usual organic layers stacked on ITO is around 5.5 eV. In order to obtain a suitable work function (around 5.3) for the ITO surface, surface treatment of the ITO is often used.

The cathode is required to have a low work function for electron injection. Stacked cathode structures such as LiF/Al, MgAg/Al, Ca/Al, or Ba/Al, are often used.

2.3 Features of the OLED

The OLED has promising potential for display and lighting applications owing to the several unique features, which are dependent upon the fact that OLEDs are solid, planar, and self-emission devices using organic materials. Table 2.1 shows how the unique features of OLEDs can produce attractive characteristics for OLED applications such as displays and lighting.

In display applications, since OLEDs have solid and planar device structures, OLEDs can realize very thin lightweight flat panel displays. Owing to the self-emission property, OLED displays can realize high contrast ratios and wide viewing angles, which are very significant factors for displays. In addition, the response time of OLEDs is as fast as micro- or nano-second order. Therefore, OLED displays can produce sharp moving images. These features are extremely attractive, compared with LCD, which is currently the major display technology. Since LCDs are non-emissive displays and utilize molecular orientational change, the levels of contrast ratio, viewing angle, response time, etc. are limited by fluctuation of molecular orientation and the restricted speed of molecular motion.

Since emission in OLEDs is caused by the emission from organic materials, emissions of various colors are possible due to the variety of organic materials. Therefore, full-color images can be created. Moreover, since white emission of OLEDs is also possible, full-color images in

Table 2.1 Features of OLEDs

Feature of OLED	Induced attractive properties in applications	
	OLED display	OLED lighting
Solid device	Thin thickness	Thin thickness
	Light weight	Light weight
Planar device	Flat panel displays	Planar lighting with minimal temperature elevation
Self-emission	High contrast ratio	Non-directive emission
	Wide viewing angle	
Fast response time	Sharp moving images	
Various emission colors	Full color	Color tunable lighting
White emission possible	Full color in combination with color filter	White lighting
Low driving voltage	TFT drive for realizing large information content	Low power consumption
	Low power consumption	
High efficiency	Low power consumption	Low power consumption

OLED displays can also be obtained by a combination with color filter. The driving voltage of OLED devices is low, such as just a few volts. Therefore, OLEDs can by driven by thin film transistors (TFT), being similar to active-matrix liquid crystal displays (AM-LCDs). The use of TFTs means that displays with high information content are possible. That is, large size TV with large information content such as full-high-vision, 4K2K or more, and high resolution mobile displays such as 500 ppi, are possible. In addition, since the efficiency of OLEDs is already very high and will be improved in the future, OLED displays can be low power consumption displays.

OLED lighting also has various attractive characteristics due to the unique feature of OLED devices.

Since OLEDs have a solid and planar device structure, they can realize thin lightweight planar lighting units. Planar lighting also has the merit of low temperature elevation because the planar shape avoids heat concentration, thereby reducing temperature elevation of the lighting.

Owing to their being self-emission devices, OLED lighting can realize emission that is non-directional.

Since white emission is also possible in OLED devices, such white OLEDs can be directly applied for lighting. In addition, when various color emissions are utilized, color tunable lighting is possible.

In addition, the high efficiency of OLED devices leads to low power consumption.

Since OLED devices utilize organic materials, lifetime and efficiency have been issues for a long time. However, with the major effort that has been applied these problems have been overcome, achieving power efficiency of almost 140 lm/W and lifetimes of 40,000 hours or more.

Another significant issue for OLEDs is cost. While the main reasons for the high cost of current OLEDs are the low utilization yield of organic materials and the large investment in production, etc., it seems that there is no reason why OLEDs should be intrinsically expensive.

In addition, it should also be emphasized that OLEDs are favorable for flexible devices. Since OLEDs are thin solid devices, they can easily be fabricated on flexible substrates. In the future, when flexible OLEDs become common, the business for OLEDs will grow significantly.

3

Light Emission Mechanism

Summary

This chapter describes the light emission mechanisms of OLEDs. There are three common types of emission mechanisms: fluorescent OLED, phosphorescent OLED, and thermally activated delayed fluorescent (TADF) OLED.

In the fluorescent OLED, only the excited singlet state gives an emission. The theoretical maximum internal quantum efficiency is only 25%. However, the phosphorescent OLED can utilize the emission from the excited singlet and triplet states, so these OLEDs can achieve an internal quantum efficiency of 100%, in principle. More recently, a third mechanism, thermally activated delayed fluorescence (TADF), was discovered, and practical materials are actively being developed.

The final section of this chapter discusses the total light emission efficiency, which is very important in actual OLED devices.

Key words

emission mechanism, fluorescent, phosphorescent, thermally activated delayed fluorescence, TADF, light emission efficiency

3.1 Fluorescent OLEDs

The most classical OLED is the fluorescent OLED. In 1987, the OLED device reported by C. W. Tang et al. [1] was a fluorescent OLED.

The emission mechanism is schematically illustrated in Fig. 3.1, showing the ground state (S_0), the singlet excited state (S_1), and the triplet excited state (T_1). By charge recombination in an organic layer, organic molecules in the emission layer are excited. In this excitation,

OLED Displays and Lighting, First Edition. Mitsuhiro Koden.
© 2017 John Wiley & Sons, Ltd. Published 2017 by John Wiley & Sons, Ltd.

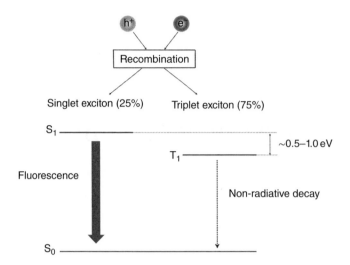

Figure 3.1   Emission mechanism of fluorescent OLED

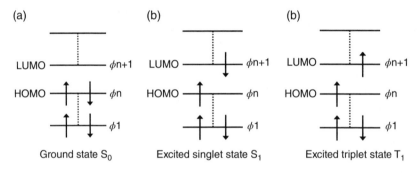

Figure 3.2   Spin conditions of the ground state (S_0), the singlet excited state (S_1), and the triplet excited state (T_1)

two types of excitons are generated: singlet and triplet excitons. The spin conditions of the ground state (S_0), the singlet excited state (S_1), and the triplet excited state (T_1) are illustrated in Fig. 3.2. The ratio of the singlet and triplet excited states is 1:3, being decided by spin statistics. The emission from the S_1 is fluorescence. The fluorescent quantum yield (Φ_{flu}) and fluorescent lifetime (τ_{flu}) are described by the following equations, where k_{flu} is a radiative rate constant, k_{nr} is a non-radiative rate constant, and k_{isc} is the intersystem crossing rate constant.

$$\Phi_{flu} = k_{flu} / \left(k_{flu} + k_{nr} + k_{isc} \right)$$

$$\tau_{flu} = 1 / \left(k_{flu} + k_{nr} + k_{isc} \right)$$

The fluorescent quantum yield (Φ_{flu}) is essentially identical to the internal quantum efficiency (IQE).

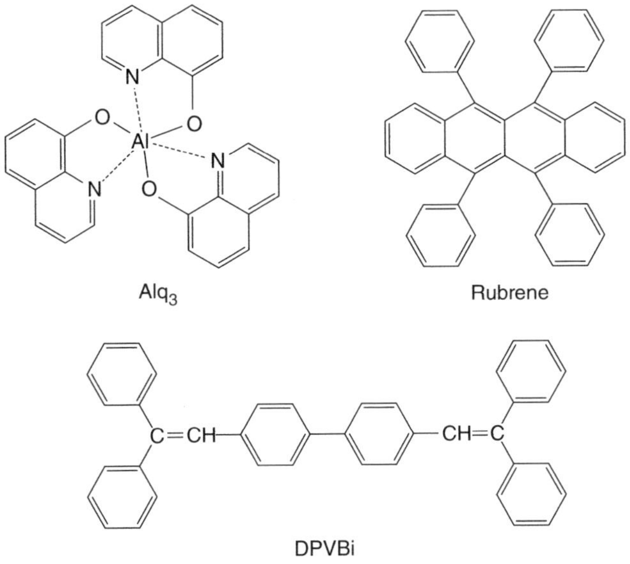

Figure 3.3 Examples of fluorescent materials

The values of τ_{flu} in OLED devices are around several nanoseconds (10^{-9}sec). The radiative rate constant k_{flu} is determined by measurements of fluorescent quantum yield (Φ_{flu}) and fluorescent lifetime (τ_{flu}), where $k_{flu} = \Phi_{flu}/\tau_{flu}$.

In fluorescent materials, only the singlet exciton can emit in competition with non-radiative decay pathways such as thermal decay. The triplet exciton does not cause an emission but thermally decays. Therefore, the maximum theoretical internal quantum efficiency of the fluorescent OLED is only 25%.

Figure 3.3 shows some typical examples of fluorescent emission materials.

3.2 Phosphorescent OLEDs

The emission mechanism of phosphorescent OLEDs is illustrated in Fig. 3.4, showing the ground state (S_0), the singlet excited state (S_1), and the triplet excited state (T_1). While the triplet state shows non-radiative decay in fluorescent materials, phosphorescent materials give an emission from not only the singlet state but also the triplet state. The well-known phosphorescent materials are complexes containing a heavy metal such as Ir, Pt, or Os, as shown in Fig. 3.5.

The triplet exciton of phosphorescent materials is changed from non-radiative to radiative by the spin-orbital coupling induced by the heavy metal effect. Therefore, the ideal internal quantum efficiency of phosphorescent OLED is 100%.

The phosphorescent OLED was reported in 1991 by T. Tsutsui et al. [2]. They discovered phosphorescence in ketocoumarin derivatives but only weak emission was observed at very low temperatures (77 K). Several years later, the group of Thompson and Forrest reported efficient phosphorescent OLED devices with 2,3,7,8,12,13,17,18-octaethyl-21H,23H-porphine platinum(II) (PtOEP) [3] and *fac* tris(2-phenylpyridine) iridium [Ir(ppy)$_3$] [4]. In addition, the internal quantum efficiency of nearly 100% has been experimentally confirmed [5, 6].

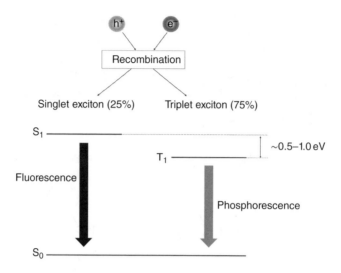

Figure 3.4 Emission mechanism of phosphorescent OLED

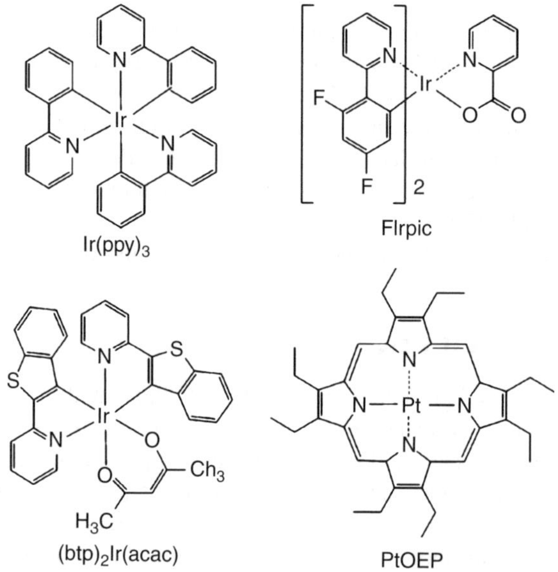

Figure 3.5 Examples of phosphorescent emission materials

Up to now, many practical phosphorescent materials have been developed and applied to commercial products. Red and green phosphorescent materials have already reached commercial level, but blue phosphorescent materials still have issues such as color purity and lifetime.

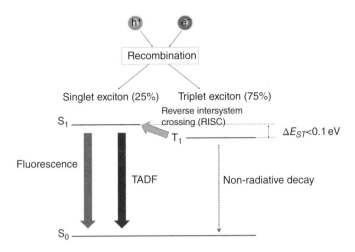

Figure 3.6 Emission mechanism of TADF OLEDs

3.3 Thermally Activated Delayed Fluorescent OLEDs

Another type of material with high efficiency is thermally activated delayed fluorescent (TADF) material [7, 8]. Figure 3.6 shows the schematic energy diagram explaining the TADF mechanism, illustrating the ground state (S_0), the singlet excited state (S_1) and the triplet excited state (T_1).

In TADF materials, the energy gap (ΔE_{ST}) between S_1 and T_1 is less than 0.1 eV. Owing to the reverse intersystem crossing (RISC) induced by the small ΔE_{ST} between S_1 and T_1, electron exchange occurs between these states. Two triplet excitons can combine to form a singlet exciton through triplet-triplet annihilation. Therefore, this small energy gap induces highly efficient spin up-conversion from the non-radiative triplet states to the radiative singlet states, giving delayed fluorescence.

The application of thermally activated delayed fluorescent (TADF) to OLEDs was reported by Adachi's group at Kyushu University [7–9]. Examples of TADF materials are shown in Fig. 3.7 [9]. The TADF phenomenon can be obtained by simple aromatic compounds, in contrast to highly efficient phosphorescent materials which require a metal complex using such rare metals as Ir, Pt, or Os.

3.4 Energy Diagram

As is described in Chapter 2 (Figs 2.1 and 2.4), plural organic layers are usually sandwiched between anode and cathode in OLED devices.

In this plural organic layer and electrodes (anode and cathode), the energy diagram is very important. An illustration of such an energy diagram is shown in Fig. 3.8.

The organic materials utilized in OLED devices are organic semiconductor materials. Therefore, each of them has two intrinsic energy levels, which are HOMO (highest occupied molecular orbital) and LUMO (lowest unoccupied molecular orbital). The energy difference between HOMO and the vacuum level is the ionization potential. The ionization potential is usually measured by photoemission spectroscopy, photoelectron yield spectroscopy, etc.

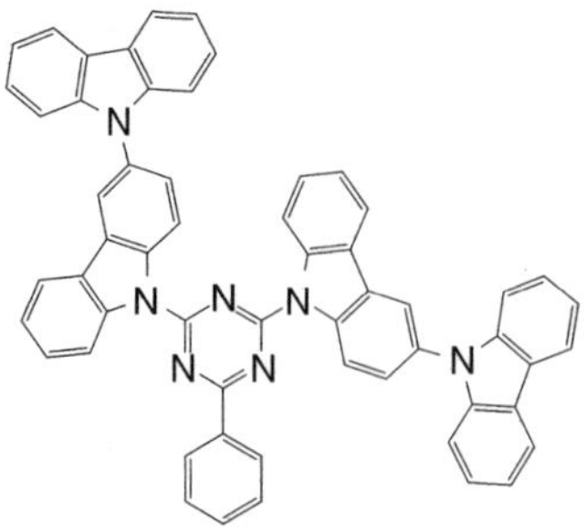

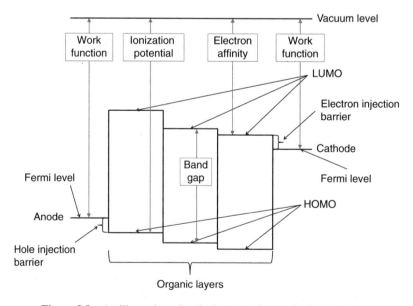

CC2TA 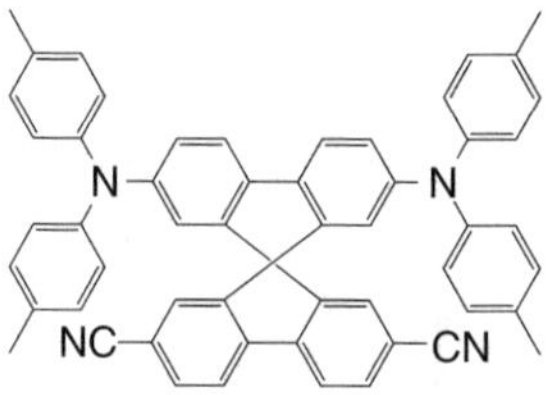   **Spiro-CN** 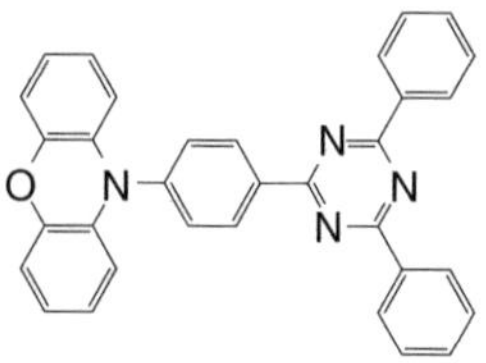   **PXZ-TRZ**

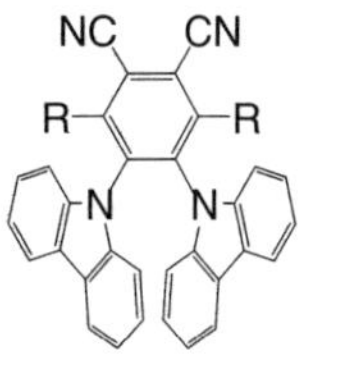

4CzPN: R = carbazolyl   **4CzIPN**   **4CzTPN: R = H**
2CzPN: R = H   **4CzN-ME: R = Me**
                                              4CzTPN-Ph: R = Me

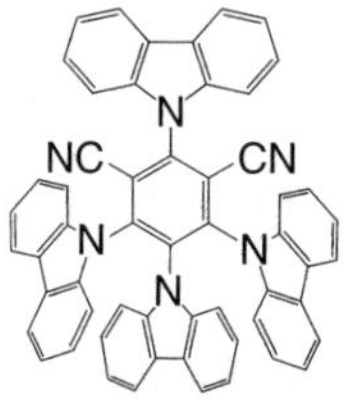

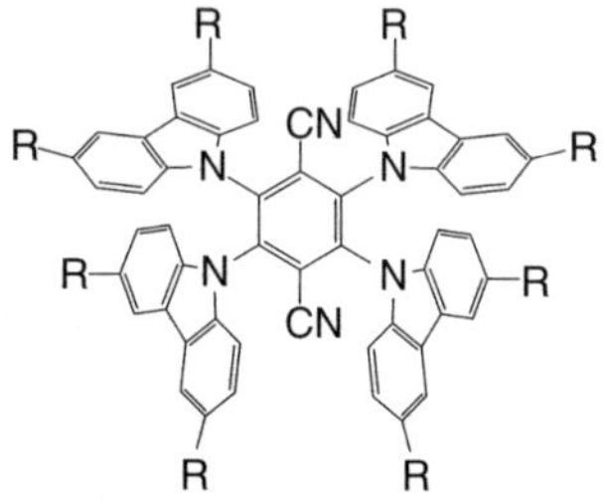

Figure 3.7 Examples of TADF-OLED materials [9]. (Copyright 2014 The Japan Society of Applied Physics)

Figure 3.8 An illustration of typical energy diagram in OLED devices

The energy difference between LOMO and the vacuum level is the electron affinity, which be measured by inverse photoelectron spectroscopy, but care needs to be taken to avoid damage caused by these measurements. The energy difference between HOMO and LUMO is the band gap. This is usually evaluated by the wavelength of the absorption edge in the absorption spectrum.

On the other hand, in electrodes, it is the Fermi level that is important. The Fermi level is a hypothetical energy level of an electron. The energy difference between the Fermi level and the vacuum level is the work function, which corresponds to the minimum energy for removing an electron.

The difference between the Fermi level of the anode and the HOMO of the organic layer adjacent to the anode is the hole injection energy barrier. This hole injection energy barrier is required to be small for in order to achieve low driving voltage.

The difference between the Fermi level of the cathode and the LUMO of the organic layer adjacent to the cathode is the electron injection energy barrier, which is also required to be small for achieving low driving voltage.

In addition, in order to achieve smooth carrier transfer to the emission layer, energy differences between adjacent layers are required to be small.

3.5 Light Emission Efficiency

The light emission efficiency of OLEDs is described by the following equation, where η_{ext} is external quantum efficiency (EQE), η_{int} is internal quantum efficiency (IQE), η_{out} is light out-coupling efficiency, γ is charge carrier balance, η_r is emissive exciton production efficiency, and q is radiative quantum yield.

$$\eta_{ext}\left(\text{external quantum efficiency}\right) = \eta_{int} \times \eta_{out} = \gamma \times \eta_r \times q \times \eta_{out}$$

This equation is schematically drawn in Fig. 3.9. The charge carrier balance γ should be balanced in order to obtain high efficient OLED devices. In other word, the number of holes is required to be same as the number of electrons, because the exceeded charge carrier means a loss of electric energy.

The emissive exciton production efficiency η_r of fluorescent OLEDs is theoretically 25%. That of phosphorescent OLED and TADF is 100% in the ideal case. It is said that the radiative quantum yield q is possible to be near 100% by using good emission materials. Therefore, if the charge carrier balance γ and radiative quantum yield q are supposed to be 100%, maximum internal quantum efficiencies η_{int} of fluorescent, phosphorescent, and TADF OLEDs are 25%, 100%, and 100%, respectively.

Another important parameter largely influencing the OLED efficiency is the out-coupling efficiency (η_{out}). If no light extraction enhancement (LEE) technology for improving the out-coupling efficiency is applied, the light out-coupling efficiency η_{out} can be roughly calculated by the reflective index difference between organic materials and air. In this case, the light out-coupling efficiency is represented by the following equation, where η_{out} is light out-coupling efficiency and n is reflective index of organic materials.

$$\eta_{out} = 1/\left(2n^2\right)$$

η_{ext} (External quantum efficiency) $= \eta_{int} \times \eta_{out} = \gamma \times \eta_r \times q \times \eta_{out}$

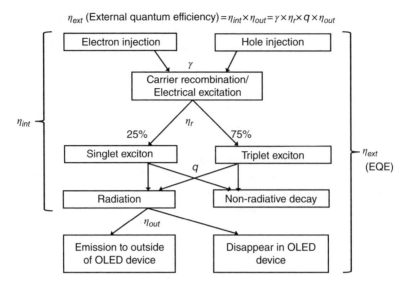

Figure 3.9   Light emission efficiency of OLEDs

Since the reflective index of common organic materials for OLEDs is around 1.6, the light out-coupling efficiency η_{out} is assumed to be about 20%. Based on this argument, the maximum external quantum efficiencies (EQEs) of fluorescent, phosphorescent, and TADF OLEDs are roughly 5%, 20%, and 20%, respectively, if no additional light extraction enhancement (LEE) technique for light out-coupling efficiency improvement is not applied.

References

[1] C. W. Tang and S. A. VanSlyke, *Appl. Phys. Lett.*, **51**, 913–915 (1987).

[2] T. Tsutsui, C. Adachi and S. Saito, *Photochemical Processes in Organized Molecular Systems*, 437–450 (1991).

[3] M. A. Baldo, D. F. O'Brien, Y. You, A. Shoustikov, S. Sibley, M. E. Thompson, and S. R. Forrest, *Nature*, **395**, 151 (1998).

[4] M. A. Baldo, S. Lamansky, P. E. Burrows, M. E. Thompson, S. R. Forrest, *Appl. Phys. Lett.*, **75**, 4–6 (1999).

[5] M. Ikai, S. Tokito, Y. Sakamoto, T. Suzuki, Y. Taga, *Appl. Phys. Lett.*, **79**, 156–158 (2001).

[6] C. Adachi, M. A. Baldo, M. E. Thompson, S. R. Forrest, *J. Appl. Phys.*, **90**, 5048–5051 (2001).

[7] A. Endo, M. Ogasawara, T. Takahashi, D. Yokoyama, Y. Kato, C. Adachi, *Adv. Mater.*, **21**, 4802 (2009).

[8] A. Endo, K. Sato, K. Yoshimura, T. Kai, A. Kawada, H. Miyazaki, and C. Adachi, *Appl. Phys. Lett.*, **98**, 083302 (2011).

[9] C. Adachi, *Jpn. J. Appl. Phys.*, **53**, 060101 (2014).

4

OLED Materials

Summary

OLED performance is largely dependent upon OLED materials. This chapter describes the classification of OLED materials and typical OLED materials.

OLED materials are divided into two types – vacuum evaporation type and solution type – from a process point of view. Vacuum evaporation materials are usually small molecular materials, while solution type materials contain polymers, dendrimers, and small molecular materials. In addition, materials are also divided into fluorescent materials, phosphorescent materials, and thermally activated delayed fluorescent (TADF) materials in terms of emission mechanisms. From the function point of view, OLED materials can be classified as hole injection material, hole transport material, emission material, host material in emissive layer, electron transport material, electron injection material, charge blocking material, etc.

Anode and cathode materials are also important, so this chapter also describes anode and cathode materials.

In addition, this chapter describes molecular orientations of organic materials because this also influences OLED characteristics.

Key words

material, vacuum evaporation, solution, small molecule, polymer, dendrimer, electrode, molecular orientation

OLED Displays and Lighting, First Edition. Mitsuhiro Koden.
© 2017 John Wiley & Sons, Ltd. Published 2017 by John Wiley & Sons, Ltd.

4.1 Types of OLED Materials

There are various types of materials for OLEDs. The classifications of OLED materials are shown in Fig. 4.1.

OLED materials are divided into two types – vacuum evaporation type (dry process) and solution type (wet process) – from a process point of view. Figure 4.2 shows schematics of the three types of OLED devices classified by process type. Hybrid type OLEDs contains organic layers deposited by solution process and those deposited by vacuum evaporation. For example, Fukagawa and Tokito et al. of NHK Science and Technical Research Laboratories (Japan) fabricated white OLED devices with solution-processed and vacuum-deposited emitting layers [1].

From a molecular structure point of view, small molecular materials, polymer materials, dendrimer materials are known in OLEDs. Vacuum evaporation materials are usually small molecular materials, while some reports have described a deposition of polymer precursors which are changed to polymers by curing on the substrate [2]. The category of solution type materials contains polymers, dendrimers, and small molecular materials.

In actual OLED devices, each material has its intrinsic function. From this point of view, OLED materials are divided to hole injection materials, hole transport materials, emitting materials, host materials in emission layer, electron transport materials, electron injection materials, and charge blocking martials, as shown in Fig. 4.1.

In addition, materials are also divided to fluorescent materials, phosphorescent materials and TADF materials from the emission mechanism point of view.

Electrode materials, which are anode and cathode materials, are also significant for OLED performance.

In actual OLED devices, various variations are possible, using materials mentioned above.

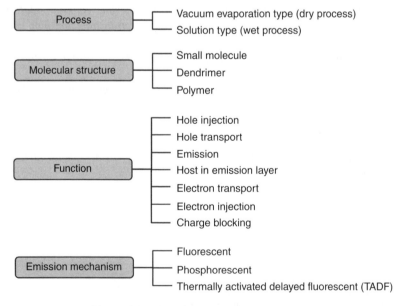

Figure 4.1 Classifications of OLED materials

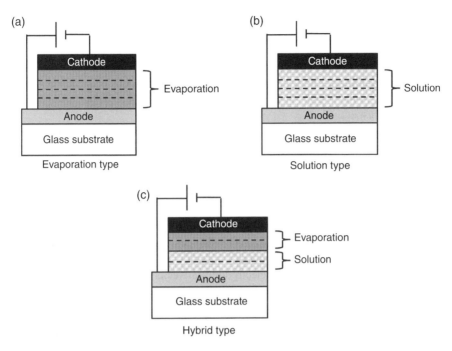

Figure 4.2 Three types of OLED devices classified by process

4.2 Anode Materials

The significant role of the anode is to inject holes into an organic layer which is adjacent to it. For this purpose, the work function of the anode is very important. Since the work function of the adjacent organic layer such as a hole injection layer (HIL) or a hole transport layer (HTL) tends to be around 5.5 eV, the work function of the anode needs to be high. In addition, normal bottom emission OLEDs require the anode to be transparent. For these reasons, ITO (indium tin oxide) has been the most popular anode material.

The ITO layer is usually fabricated by sputtering deposition. The work function of ITO is about 4.5–5.2 eV, depending on the ITO fabrication process, film properties, surface condition, etc. Just before the next deposition of the adjacent organic layer, surface of the ITO is often treated by O_2 plasma [3, 4] or UV-O_3 [4, 5] for the purpose of eliminating any organic materials on the surface. Such surface treatments also tend to give rise to an increase in work function, giving an improvement in the hole injection property.

Milliron et al. of Princeton University reported that oxygen plasma treatment of ITO increases its work function by about 0.5 eV [3]. Also, Wu et al. in Princeton University investigated about the effect of O_2 plasma, comparing it with other treatment methods. In addition, they found a lack of significant change in surface morphology, indicating that morphology is not the dominant factor in the improvement of device performance and reliability. It was confirmed that UV ozone and oxygen plasma treatments have an effect on turn-on voltage and efficiency. In particular, they reported that oxygen plasma treatments show remarkable effects. On the other hand, Ar and H_2 plasma treatments gave no positive influence. They suggested that change in the surface chemical composition might play a role in increasing the hole injection ability at the ITO/organic interface.

Other surface treatment methods have also been studied and reported. In the first, Hung et al. of Eastman Kodak Company reported on anode modification by low-frequency plasma polymerization of CHF_3 [6]. The thickness of polymerized film ranges from 2.5 to 10 nm. They reported that enhanced hole injection and superior operational stability were obtained. For example, the luminance drop in the first 150 h under continuous operation at 40 mA/cm^2 was 1% and 15% in the OLED device with a 6 nm thick polymerized buffer layer and later without a buffer, respectively.

Ganzorig et al. of Fujihara's group at the Tokyo Institute of Technology (Japan) studied and reported about the surface treatment method with benzoyl chlorides [7]. They chemically modified the ITO surface with H-, Cl-, and CF_3-terminated benzoyl chlorides and fabricated OLED devices using the modified ITO. They observed a remarkable reduction in driving voltage, due to the enormous increase in ITO work function. In addition, they found that the order of reduction in drive voltage is consistent with the order of the magnitude of the permanent dipole moments ($\mu_{CF3-} > \mu_{Cl} > \mu_{H-}$).

Applications of self-assembled monolayers (SAMs) have also been reported. Campbell et al. of Los Alamos National Laboratory and The University of Texas and Dallas (USA) reported that the application of SAM on metal electrode improves charge injection, manipulating the Schottky energy barrier between a metal electrode and the organic material [8].

Recently, due to such problems of ITO as frigidity, rare metal issues, and cost, active research and development on non-ITO transparent electrodes has started, aimed at the application to not only OLEDs but also touch-panels, etc. Such non-ITO transparent electrodes will be described in Section 11.1.

Top emitting OLEDs, on the other hand, require a reflective anode. Materials such as Ag, Ag/ITO, and ITO/Ag/ITO, are often used as the anode for top emitting OLEDs.

Ag is a strong candidate due to its high reflectivity for visible light and low electrical resistivity. However, since the work function of Ag is about 4.3 eV, a large hole injection barrier remains. Therefore, Ag itself is not adequate for hole injection.

The easiest way seems Ag/ITO stacking. Hsu et al. of National Chiao Tung University and National Tsing Hua University (Taiwan, Republic of China) reported top emitting OLED devices with Ag/ITO stacking [9].

Chen et al. of National Taiwan University (Taiwan, Republic of China) reported that the introduction of a thin Ag_2O layer (around 10 nm) on Ag gave good hole injection, which resulted in good OLED characteristics [10]. Ag_2O is known to show p-type semiconductor property with a Fermi level of 4.8–5.1 eV. The group made thin Ag_2O by UV-ozone treatment of Ag, retaining a high reflectivity of 82–91%, which is only slightly lower than that of an as-deposited Ag film.

Chong et al. of National Cheng Kung University (Taiwan, Republic of China) reported on the modification of silver anode by an organic solvent (tetrahydrofuran) [11]. After deposition of a silver film on ITO coated glass, the Ag anode was immediately treated by immersing it in THF for 30 minutes. After the reaction, the THF-modified Ag anode was dried with a nitrogen steam and delivered to the nitrogen-filled glove-box. The X-ray photoelectron spectroscope analysis shows that the THF molecules were chemically adsorbed on the Ag surface, forming oxygen-rich species by substrate-catalyzed decomposition. The work function of the THF-modified Ag is 4.79 eV, while that of the base Ag was 4.54 eV. On this substrate, a top emitting OLED device with the structure of ITO/THF-modified Ag/HY-PPV/Ca(12 nm)/Ag(17 nm) was fabricated. The HY-PPV is phenyl-substituted poly(para-phenylene-vinylene) copolymer.

The device showed 2.93 cd/A, while the reference device with non-modified Ag showed only 0.51 cd/A. This suggests that the increase in the current efficiency is attributed to the increase in the work function induced by the THF treatment.

4.3   Evaporated Organic Materials (Small Molecular Materials)

Vacuum evaporation materials are most widely used in current OLED technologies which include commercial products. In general, the materials are small molecular materials and are divided to hole injection materials, hole transport materials, emitting materials, host materials in emission layer, electron transport materials, electron injection materials, charge blocking materials, etc. by their function. The following subsection describes these materials.

4.3.1   Hole Injection Materials

In OLED devices, there is an energy barrier between the anode and the adjacent organic layer. Therefore, it is required to reduce this barrier for obtaining smooth hole injection from the anode to the adjacent organic layer. For this purpose, hole injection materials are often used in practical OLED devices because they can contribute to the reduction in drive voltage, the increase in efficiency, the elongation of lifetime, etc.

Typical hole injection materials are shown in Fig. 4.3. They are divided to organic materials and inorganic materials, while some hybrid materials are also possible.

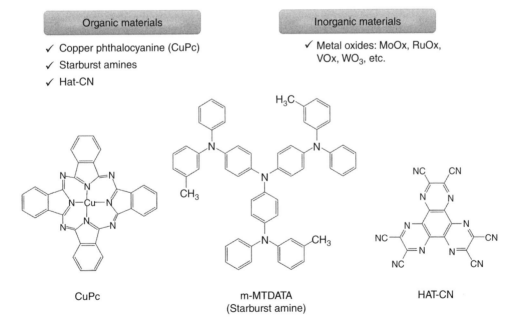

Figure 4.3   Typical hole injection materials

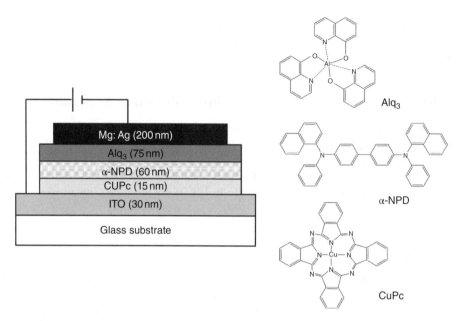

Figure 4.4 Device structure and molecular structures of utilized materials reported by VanSlyke et al. [12]

Copper phthalocyanine (CuPc) is one of the most well-known hole injection materials. VanSlyke et al. of Eastman Kodak reported an insertion of CuPc layer between an ITO anode and a hole transport layer of N,N'-diphenyl-N,N'-bis(1-naphthyl)-1,10-biphenyl-4,4'-diamine (α-NPD) [12]. The reported device structure and the molecular structures of utilized materials are shown in Fig. 4.4. The ionization potential of CuPc is 4.7 eV, being lower than that of α-NPD (5.1 eV). Since the ionization potential of ITO is around 4.7 eV, the ITO/CuPc barrier is lower than the ITO/α-NPD barrier. They reported that such lowering of the interfacial barrier can affect the hole injection efficiency and can contribute to the improvement of lifetime.

Starburst amines with bulky molecular structures are also often used as hole injection materials. They were firstly reported by Shirota et al. of Osaka University [13, 14]. The typical material is 4,4',4''-tris{N,(3-methylphenyl)-N-phenylamino}-triphenylamine) (m-MTDATA) shown in Fig. 4.3.

Shirota et al. reported that the insertion of m-MTDATA gives rise to an increase in efficiency and lifetime, investigating the influence of m-MTDATA on OLED performances. The OLED device with the structure of ITO/m-MTDATA(60 nm)/TPD(10 nm)/Alq$_3$(50 nm)/MgAg is reported to show about 30% higher quantum efficiency and longer lifetime than the reference device with the structure of ITO/TPD(10 nm)/Alq$_3$(50 nm)/MgAg [14].

Another typical HIL material is 1,4,5,8,9,11-hexaazatriphenylene-hexacarbonitrile (HAT-CN). The molecular structure is shown in Fig. 4.3. HAT-CN is often used in practical OLED devices because of the attractive properties such as high mobility and excellent hole injection to the adjacent HTL layer. Rhee et al. of Sunmoon University and Korea Institute of Materials Science (South Korea) have investigated the effect of HAT-CN, comparing it with other HIL materials [15].

Some inorganic oxides have been reported to be applied as a hole injection layer, improving the OLED performance. In particular, molybdenum trioxide (MoO_3) is often used in practical OLED devices.

Tokito et al. of Toyota Central Research and Development Laboratories (Japan) have reported on applying such thin metal oxides as vanadium oxide (VO_x), molybdenum oxide (MoO_x) and ruthenium oxide (RuO_x) as a hole injection layer [16]. By using such metal oxides for hole injection, the operating voltage is reduced with respect to a device with the more common ITO electrode. This effect is attributed to the reduction in energy barrier for hole injection to the hole transport layer.

Other metal oxides such as SiO_2 [17], CuO_x [18], NiO [19, 20], and WO_3 [21], have also been proposed and investigated as hole injecting materials.

Matsushima and Murata et al. of Japan Advanced Institute of Science and Technology investigated the effect of the thickness of MoO_3 on the hole injection properties, finding that the OLED device with a 0.75 nm MoO_3 layer forms ohmic hole injection at the ITO/MoO_3/α-NPD interfaces and I–V characteristics of this device are controlled by a space-charge-limited current [22].

Another approach is to use a p-doped hole injection layer. Zhou and Leo et al. of Technische Universität Dresden (Germany) reported on a hole injection layer consisting of TDATA doped with F_4-TCNQ [23]. The device structure and molecular structure of TDATA is shown in Fig. 4.5. They reported that the operating voltage for luminance of $100 \, cd/m^2$ is 3.4 V in an OLED device with 2 mole % of F_4-TCNQ, while it is 9 V in an OLED without F_4-TCNQ.

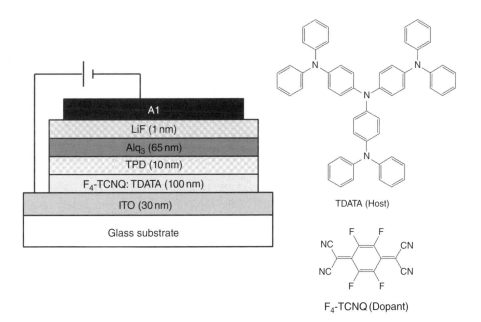

TDATA (Host)

F_4-TCNQ (Dopant)

Figure 4.5 Device structure of OLED with a hole injection layer consisting of TDATA doped with F_4-TCNQ [23]

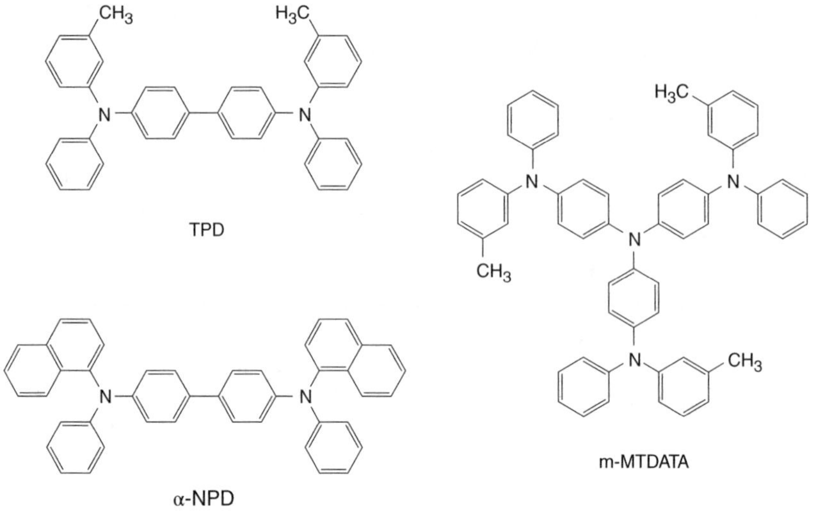

Figure 4.6   Typical hole transport materials for OLEDs

4.3.2   Hole Transport Materials

The role of hole transport materials is to transport holes to emission layers. Typical hole transport materials for OLEDs are aromatic amino compounds as shown in Fig. 4.6.

One of classical hole transport materials was N,N'-diphenyl-NN'-bis(3-methylphenyl)-[1,1'-biphenyl]-4,4'-diamine (TPD), which was used in the OLED device reported by Tang et al. in 1987 [24]. TPD has high hole mobility and easily gives an amorphous film on substrates by vacuum deposition. However, the film has the demerit of crystallization on long-term storage at room temperature, due to the low glass transition temperature (Tg about 60 °C). It is known that this crystallization reduces the effective contact area with the anode due to the change in film structure, giving a drastic reduction of efficiency [25].

As hole transport materials with improved glassy temperature, two notable materials have been developed: 4,4'-bis[N-(1-naphthyl)-N-phenyl-amino]biphenyl (α-NPD) and starburst amines.

The glass transition temperature (Tg) of α-NPD is about 95 °C [26]. The oxidation of the natural α-NPD molecule and the reduction of the α-NPD$^+$ cation are completely reversible. In addition, the double charged α-NPD^{++} cation is similarly reversible in oxidation/reduction cycle.

Starburst amines [13,14,27–29], which are also used as hole injection materials, can be used as hole transport materials. Typical starburst amines are shown in Fig. 4.7. These materials have bulky molecular structures and prohibit a planar shape, preventing reorientation of the molecules and consequently crystallization. Because of this effect, starburst amines tend to easily give a stable amorphous glassy state which is required in OLED devices. Starburst amines are also used as hole injection materials in combination with other hole transport materials such as α-NPD [14].

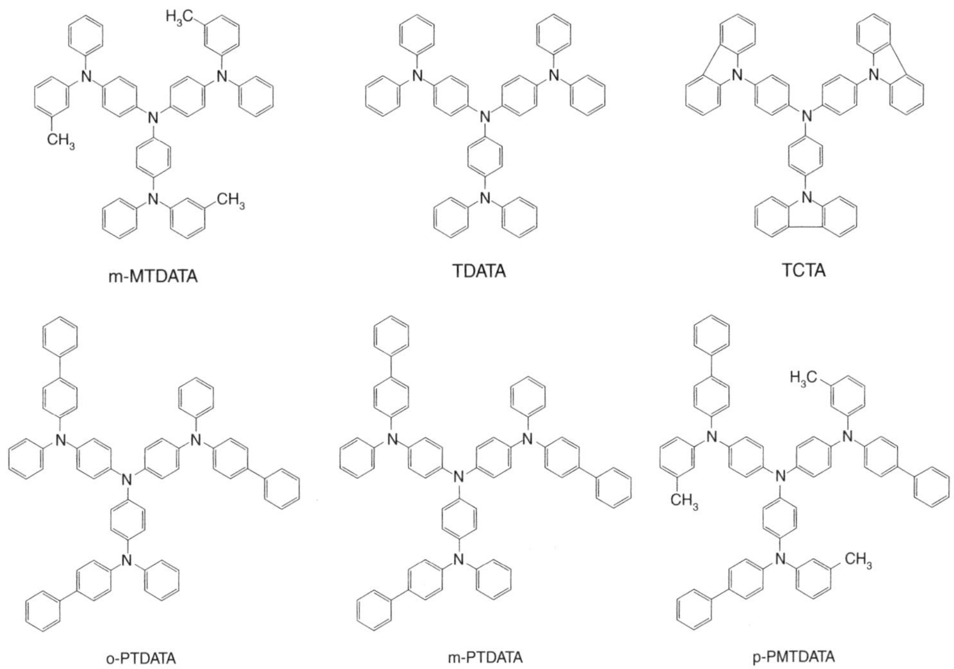

Figure 4.7 Typical starburst amines for OLEDs

4.3.3 Emitting Materials and Host Materials in Fluorescent Emission Layer

The most significant role of materials in the emission layer of OLEDs is to give highly efficient emission and the required color. Indeed, emitting materials are strongly related to the emitting efficiency, emitting color, lifetime, etc. Materials in emitting layers are often composed of host and dopant, which is also referred to as a guest–host system. Some examples of emitting and host materials for fluorescent emitting layers are shown in Fig. 4.8.

Typical emitting dopants are perylene (blue), distyrylamine (blue) [30], coumarin (green) [31], Alq_3 (green), rubrene (yellow) [32], and dicyano methylene piran derivatives (red) such as DCM, DCM-1, DCM-2, DCJTB [31–33]. As host materials, Alq_3, distyrylarylene (DSA), etc. are well known. Alq_3 is a green-emitting material by itself, and is also often used as a host material.

In such guest–host systems, the energy gap of the guest dopant should be smaller than that of the host material. The doping concentration is usually 0.5–5%. A higher concentration is not suitable because it induces concentration quenching.

At present, blue-emitting fluorescent OLED materials are still usually used in commercial OLED products because of the absence of commercial level deep blue phosphorescent materials, while commonly used green and red materials have already been changed to phosphorescent OLED materials.

Hosokawa et al. of Idemitsu Kosan (Japan) reported blue OLED devices consisting of a DSA host with an amino-substituted DSA dopant [30]. Examples of DSA host and

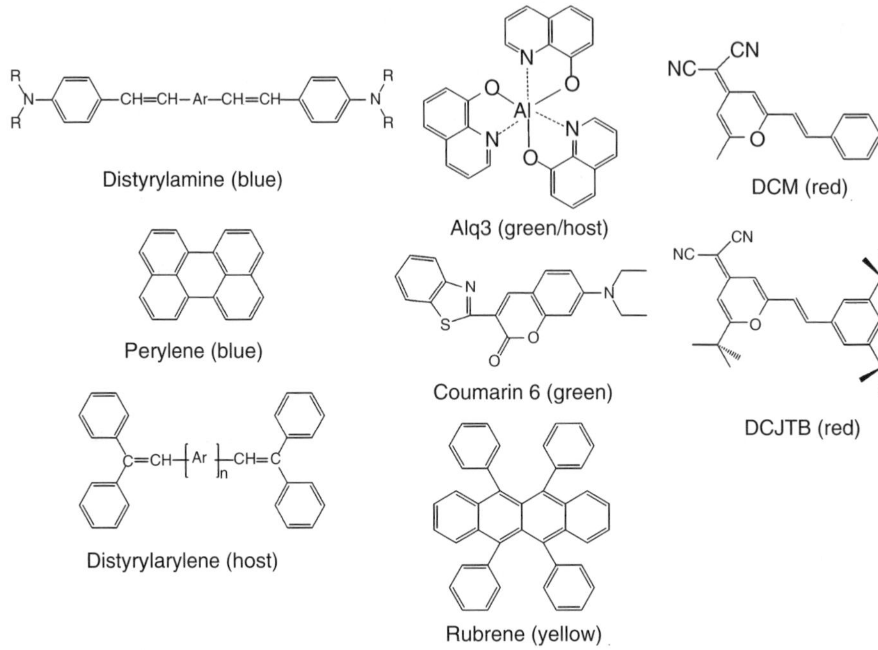

Figure 4.8 Examples of emitting and host materials for fluorescent emitting layers

amino-substituted DSA dopant are shown in Fig. 4.9. They reported, in 1997, achieving 6 lm/W and a half lifetime of 20,000 hours at an initial luminance of 100 cd/m^2, by using a DSA emitting layer containing a small amount of an amino-substituted DSA dopant [34]. These materials have been the basis of first the commercial blue-emitting OLED products.

One interesting phenomenon in fluorescent OLED materials is triplet-triplet fusion (TTF). Kawamura et al. of Idemitsu Kosan (Japan) reported that triplet excitons (^{3}A*) collide with each other and generate singlet excitons (^{1}A*) by the following formula [35]:

$$^3A* + {}^3A* \rightarrow (4/9)^1 A + (1/9)^1 A* + (13/9)^3 A*$$

They call this phenomenon triplet-triplet fusion (TTF). Using the TTF phenomenon, Kawamura et al. reported the development of a deep blue fluorescent OLED with CIEy of 0.11 and high EQE of over 10% [36]. They also developed a bottom emission OLED showing a CIE1931 coordinate of (0.14, 0.08) and a current efficiency of 6.5 cd/A [35].

4.3.4 Emitting Materials and Host Materials in Phosphorescent Emission Layer

After the first report on efficient phosphorescent OLED devices at room temperature by Baldo et al. of Thompson and Forrest's group at Princeton University (USA) [37, 38], aggressive developments on phosphorescent OLEDs started.

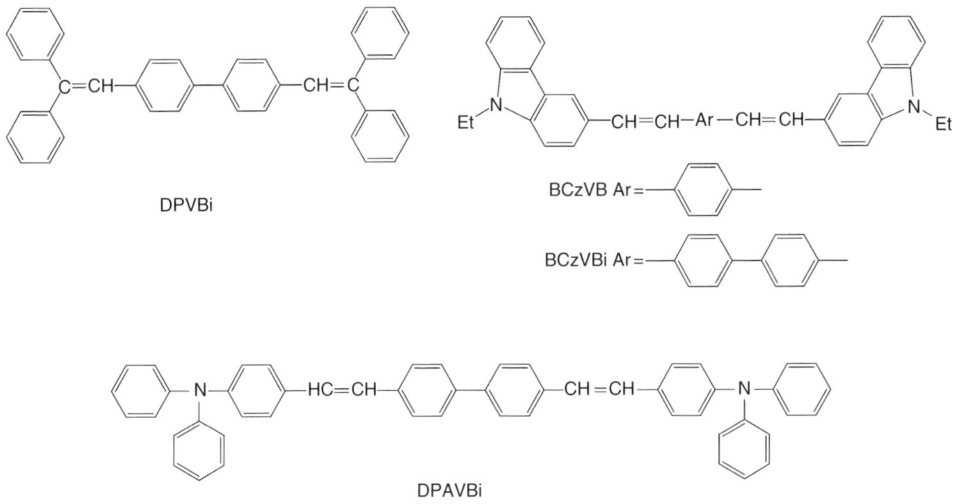

Figure 4.9 Molecular structures of distyrylarylene (DSA) derivatives in blue OLED devices reported by Hosokawa et al. [30]

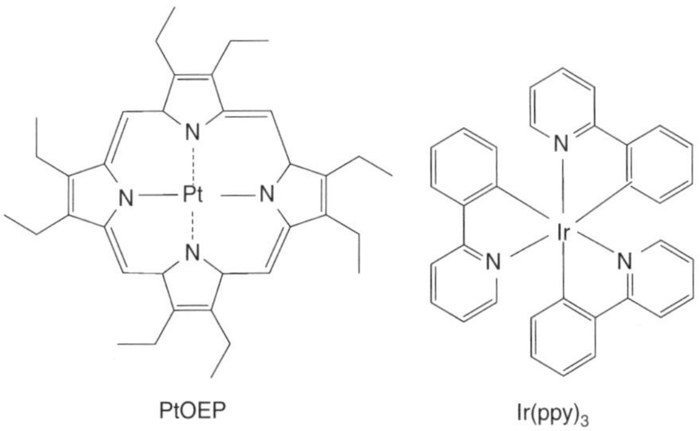

Figure 4.10 Molecular structures of PEOEP and Ir(ppy)₃

In their first report, they used 2,3,7,8,12,13,17,18-octaethyl-21H,23H-porphine platinum(II) (PtOEP) as a phosphorescent emitter. The molecular structure of PtOEP is shown in Fig. 4.10. While they observed a phosphorescent emission, they obtained maximum EQE of only 4% [37]. In their next report, they utilized iridium phenylpyridine complex Ir(ppy)$_3$ (Fig. 4.10) and reported high efficiencies such as an EQE of 8%, current efficiency of 28 cd/A, and power efficiency of 31 lm/W. [38].

Figure 4.11 shows the energy diagram of the OLED device with Ir(ppy)$_3$ [38]. In the emission layer, a few % of Ir(ppy)$_3$ was doped to the host material CBP. In their experiments, when the doping concentration was 6%, the maximum efficiency was obtained. A thin barrier layer of BCP was inserted between the emission layer and the electron transporting layer (Alq$_3$) for confining excitons within the luminescent zone and hence maintaining high efficiencies.

In addition, internal quantum efficiency of nearly 100% has been experimentally confirmed [39–41].

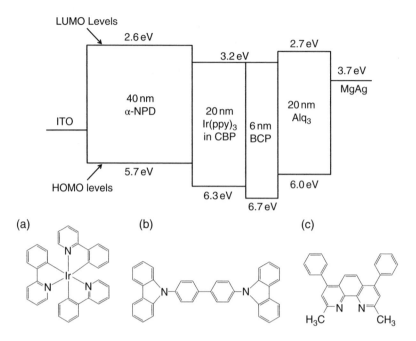

Figure 4.11 Proposed energy level structure of the the OLED device with Ir(ppy)$_3$ in the report of Baldo et al. [38]. Note that the HOMO and LUMO levels for Ir(ppy)$_3$ are unknown. The inset shows the chemical structural formulas of (a) Ir(ppy)$_3$, (b) CBP, and (c) BCP

4.3.4.1 Phosphorescent Emitting Dopants

As described above, emission layers usually consist of phosphorescent dopant and host material in phosphorescent OLED devices. Emitters for phosphorescent OLEDs are metal complexes that have a heavy metal such as iridium (Ir), platinum (Pt), ruthenium (Ru), osmium (Os), or rhenium (Re).

The most frequently used emitting dopants for phosphorescent OLED are Ir-complexes. Examples of phosphorescent Ir-complexes are shown in Figs 4.12 and 4.13. While various Ir-complexes have been reported, they can be divided to tris-ligandated type (Fig. 4.12) and di-ligandated type (Fig. 4.13). One of the most famous Ir-complexes for phosphorescent OLEDs is Ir(ppy)$_3$. This material gives green emission. By changing the ligand, various colors can be obtained [42, 43], while the efficiency, lifetime, etc. are also largely dependent upon the ligands. The emission is assigned as MLCT (metal to ligand charge transfer).

It is known that there are two geometrical isomers, which are facial and meridional isomers, as shown in Fig. 4.14. Tamayo et al. of University of Southern California (USA) investigated about facial and meridional isomers of several Ir-complexes [44] and reported the following:

1. The facial isomer is thermodynamically stable in comparison with the meridional isomer. The facial isomer is the thermodynamic product and the meridional isomer is kinetic product. The meridional isomer is easily transferred to the facial isomer at high temperatures.
2. Selective synthesis of each isomer is possible by controlling the reaction condition.

$Ir(F_2ppy)_3$ [Blue]

$Ir(ppy)_3$ [Green]

$Ir(bzq)_3$ [Yellow]

$Ir(pic)_3$ [Red]

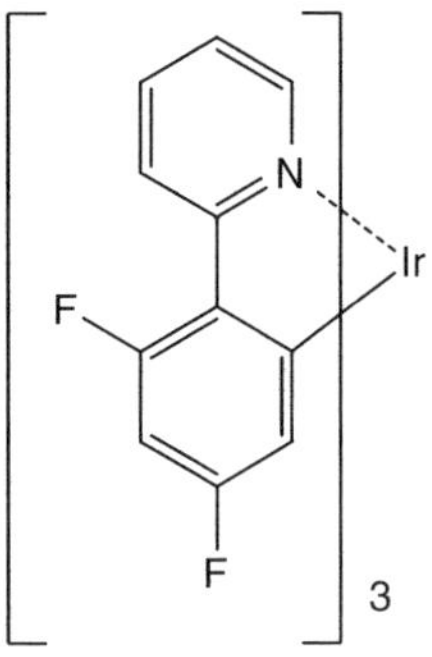

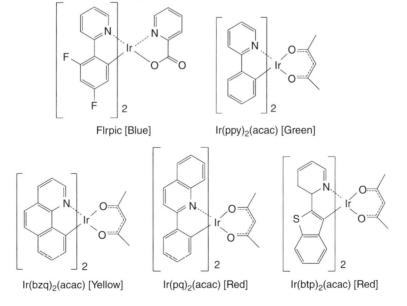

Figure 4.12 Tris-liganded type of phosphorescent Ir complexes

FIrpic [Blue]

$Ir(ppy)_2(acac)$ [Green]

$Ir(bzq)_2(acac)$ [Yellow]

$Ir(pq)_2(acac)$ [Red]

$Ir(btp)_2(acac)$ [Red]

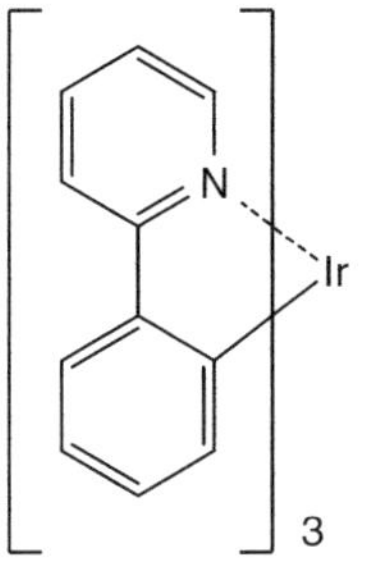

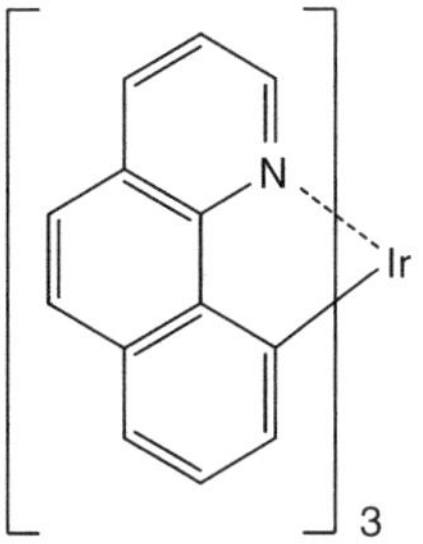

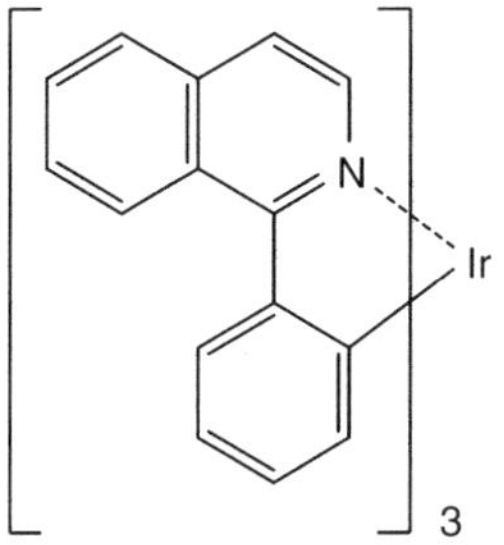

Figure 4.13 Di-liganded type of phosphorescent Ir complexes

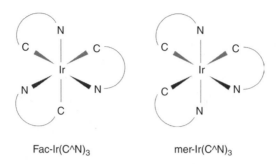

Fac-Ir(C^N)$_3$ mer-Ir(C^N)$_3$

Figure 4.14 Facial and meridional isomers of Ir(ppy)$_3$

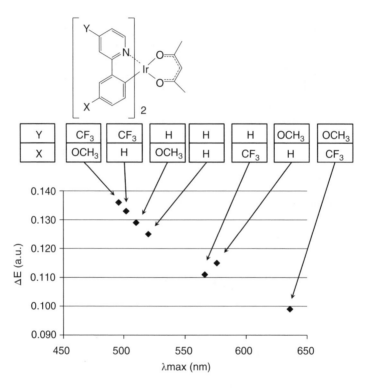

Figure 4.15 The substituent effect of phosphorescent Ir-complexes on the emission wavelength [45]. ΔE is the calculated energy difference between the HOMO and LUMO levels. λmax is the photoluminescent peak wavelength of the solution containing each phosphorescent Ir complex

3. The meridional isomer shows broad and red-shifted emission relative to the facial isomer.
4. The photoluminescent quantum efficiency of the facial isomer is higher than that of the meridional isomer.

Yoshihara et al. of Gunma University (Japan) reported on the substituent effect of phosphorescent Ir-complexes on the emission wavelength [45]. The results are shown in Fig. 4.15.

The spectra of PL phosphorescent emissions were measured in toluene solution. They calculated the energy difference ΔE between the HOMO and LUMO levels. It is found that the emission peak wavelength is almost correlated with the ΔE. In the Y-position, the electron withdrawing CF_3 group induces a decrease in the emission wavelength but the electron releasing OCH_3 group induces an increase in the emission wavelength. On the contrary, the relationship at the X-position is opposite. That is, in the X-position, the electron withdrawing CF_3 group induces an increase in the emission wavelength but the electron releasing OCH_3 group induces a decrease in the emission wavelength.

At present, while red and green phosphorescent materials have been applied to commercial products, blue phosphorescent materials still have problems especially in color purity and lifetime.

For blue phosphorescent materials, FIrpic (iridium(III)bis(4,6-di-fluorophenyl)-pyridinato-N,C-2′)picolinate) and FIr6 (iridium(III)bis(4′,6′-difluorophenylpyridinato)tetrakis(1-pyrazolyl)borate) are famous emitters, showing an emission peak at 470 and 458 nm, respectively. However, both emission spectra show a long tail into the green spectral region which shifts into the bluish green color region. Therefore, blue phosphorescent emitters are still being actively developed.

Some examples of newly developed blue phosphorescent emitters are shown in Fig. 4.16.

A series of cyclometallated carbene iridium complexes were promising candidates for a blue emitter with high η_{PL} [46]. Typical material is *mer*-tris(N-dibenzofuranyl- N′-methylimidazole)iridium(III) [Ir(dbfmi)] showing blue emission with a λmax at 445 nm. The blue OLED with Ir(dbfmi) showed a maximum power efficiency of 35.9 lm/W. Using this martials, a white OLED device with a maximum power efficiency of 59.9 lm/W was achieved without any light-out-coupling enhancement.

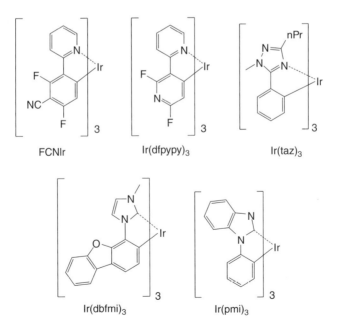

Figure 4.16   Several blue phosphorescent emitters

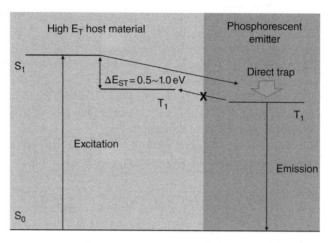

Figure 4.17 Energy transfer model of phosphorescent OLED with host material and phosphorescent emitter

4.3.4.2 Host Materials for Blue Phosphorescent OLEDs

While red and green phosphorescent OLED materials have been widely used in commercial products, blue phosphorescent OLED materials still have problems. One of the reasons is the lack of a suitable host material for phosphorescent blue emission layer.

Since blue phosphorescent OLEDs require wide energy gap organic materials or high triplet excited energy level (E_T) materials, organic host materials and charge transporting materials having the wide energy gap need to be developed as shown in Fig. 4.17. In addition, the compatibility of small ΔE_{ST} (energy gap between S1 and T1 states) and high E_T is necessarily required. In blue phosphorescent OLEDs, host materials with high triplet energy (E_T) of over 2.75 eV is required.

Adachi et al. of Forrest and Thompson's group at Princeton University (USA) reported that an OLED device with FIrpic doped to 4,4′-N,N-′-dicarbazole-biphenyl (CBP) showed blue phosphorescent emission with the peak wavelength of 470 nm, a very high maximum organic light-emitting device external quantum efficiency of $(5.7 \pm 0.3)\%$ and a luminous power efficiency of (6.3 ± 0.3) lm/W [42]. However, the triplet energies of CBP and FIrpic are 2.6 eV and 2.7 eV, respectively [47,48]. Since the triplet energy of CBP is lower than FIrpic, the energy confinement is not sufficient.

To solve this problem, Tokito et al. of NHK Science and Technical Research Laboratories (Japan) utilized CDBP (4,4′-bis(9-carbazolyl)-2,2′-dimethyl-biphenyl), which is a dimethyl substituted derivative of CBP [47, 48]. The molecular structures and energy levels of CDBP, CBP and FIrpic are shown in Fig. 4.18. The triplet energy of CDBP is 3.0 eV, which is higher than CBP and FIrpic. The increase in the triplet energy of CDBP seems to be attributed to the kink of biphenyl moiety, which is derived from the introduction of two methyl groups at the ortho position of the biphenyl. Since the triplet energy of CDBP is larger than FIrpic, the effective energy transfer from the CDBP triplet state to the FIrpic triplet state seems to occur, giving high efficiency.

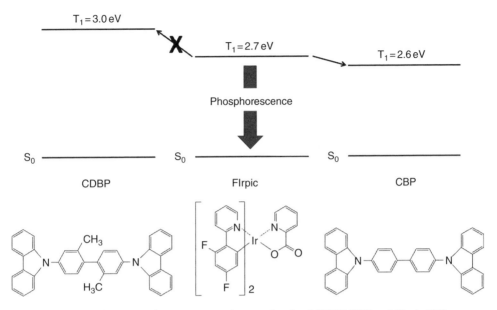

Figure 4.18   Molecular structures and energy levels of CDBP, CBP, and FIrpic [47]

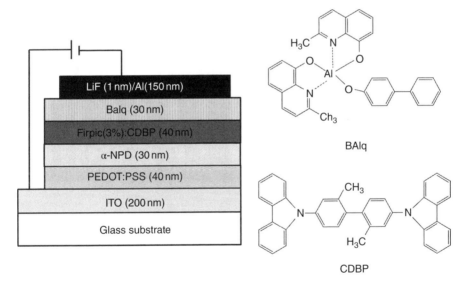

Figure 4.19   A blue phosphorescent device reported by Tokito et al. [47]

Indeed, Tokito et al. developed a phosphorescent OLED device with FIrpic and CDBP, obtaining high efficiency [47]. The device structure and some utilized materials are shown in Fig. 4.19. They reported achieving a maximum external quantum efficiency of 10.4%, which corresponds to a current efficiency of 20.4 cd/A.

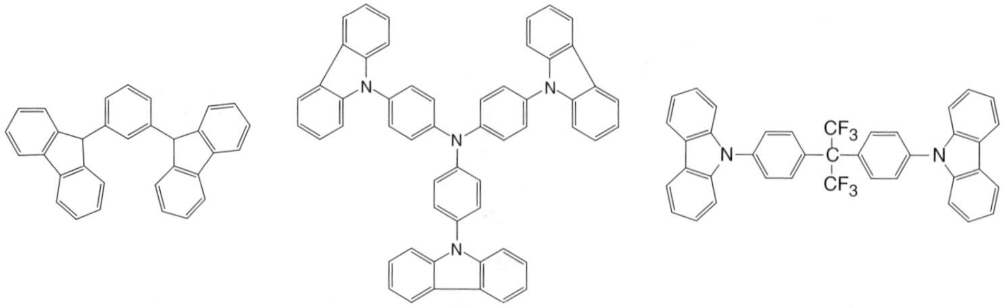

Figure 4.20 Example of recently developed host materials for blue phosphorescent OLEDs

Thus far, various types of host materials for blue phosphorescent OLEDs have been investigated and developed. Some examples are shown in Fig. 4.20. Sasabe et al. in Yamagata University (Japan) developed a highly efficient multi-photon emission blue phosphorescent OLED, achieving 90 cd/A and 41 lm/W [49].

4.3.5   Emitting Materials and Host Materials in TADF Emission Layers

Thermally activated delayed fluorescent (TADF) materials are attractive next generation materials because TADF can achieve high efficiency even with fluorescent materials with no rare metal. As described in Section 3.3, TADF materials require small energy gap (ΔE_{ST}) between the excited singlet state (S_1) and the excited triplet state (T_1) as shown in Fig. 3.6. In commonly used fluorescent materials, the ΔE_{ST} is assumed to be 0.5–1.0 eV. Uoyama et al. of Adachi's group at Kyushu University (Japan) described that the critical point of the molecular design in highly luminescent TADF materials is a compatibility of a smaller ΔE_{ST} than 0.1 eV and a longer radiative decay rate than 10^6/s to overcome competitive non-radiative pathways [50]. Since these two requirements conflict with each other, it is required that the overlap of the highest occupied molecular orbital (HOMO) and the lowest unoccupied molecular orbital (LUMO) is carefully balanced. In addition, they also commented that the geometrical change in molecular orientation between the S_0 and S_1 states should be restricted to suppress non-radiative decay.

Uoyama et al. reported on the design of a series of highly efficient TADF emitters based on carbazolyl dicyanobenzene (CDCB) [50]. Examples of TADF emitters with the CDCB structure are shown in Fig. 4.21. The carbazole unit plays the role of electron donor, and the dicyanobenzene acts as electron accepter. The series of CDCB materials shows a wide range of emission colors ranging from sky blue with an emission peak of 473 nm to orange with an emission peak of 577 nm. The emission wavelength depends on the electron-donating and electron-accepting abilities of the peripheral carbazolyl group and the central dicyanobenzene unit, respectively. They fabricated OLED devices with the TADF materials as shown in Fig. 4.22. The green OLED device is reported to show a very high EQE of 19.3 ± 1.5%, which is equivalent to an internal EQE of 64.3–96.5% assuming a light out-coupling efficiency of 20–30%.

Further active developments are being done of obtain practically useful materials [51–53]. Some examples are shown in Fig. 4.23.

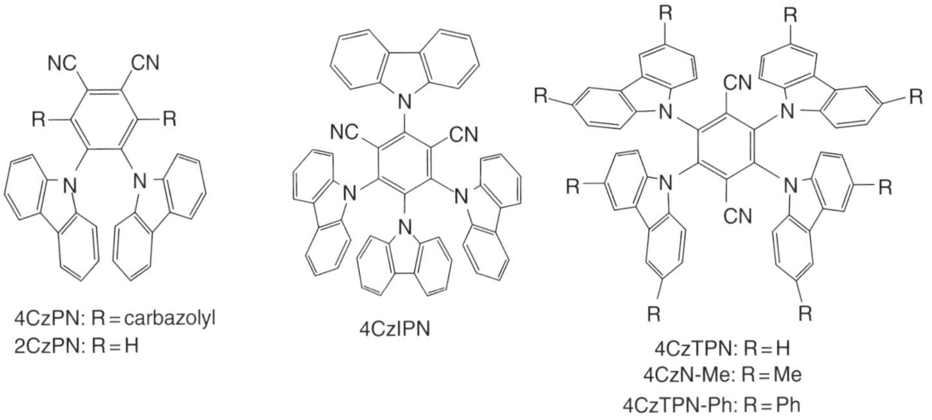

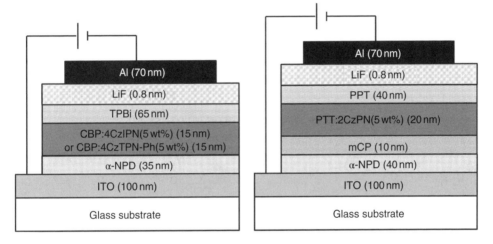

Figure 4.21   Examples of TADF emitters with CBCB structure [50]

Figure 4.22   OLED devices with the TADF materials [50]

4.3.6   Electron Transport Materials

The roll of electron transport materials is to bring injected electrons to the emitting layers. Some examples of classical electron transport materials are shown in Fig. 4.24. One of the most famous electron transport materials is tris(8-hydroxyquinoline)aluminum (Alq_3), which also acts as a green emitter. Since Alq_3 was applied to OLED devices, as reported by Tang and VanSlyke [54], Alq_3 has often been used as not only an emitting material but also an electron transport material.

However, OLED devices with Alq_3 tend to show high drive voltage and low efficiency, due to its low electron mobility and injection properties. Based on such knowledge, various electron transport materials have been developed, aiming at the improvement of OLED performance. In these developments, electron-deficient aromatic moieties such as oxadiazoles, triazoles, imidazoles, pyridines, and pyrimidines, have often been used as building blocks for electron transport materials to accept electrons from the cathode efficiently.

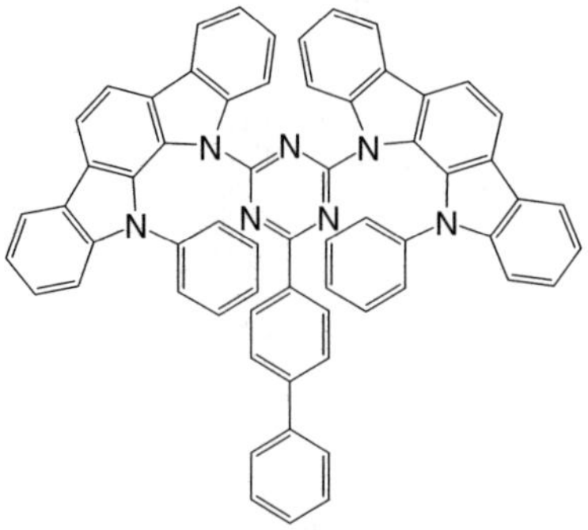

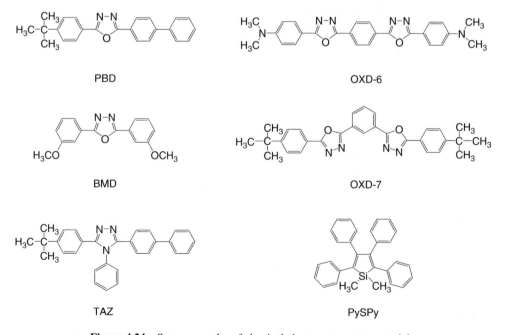

Figure 4.23   Examples of TADF emitters [51–53]

PIC-TRZ

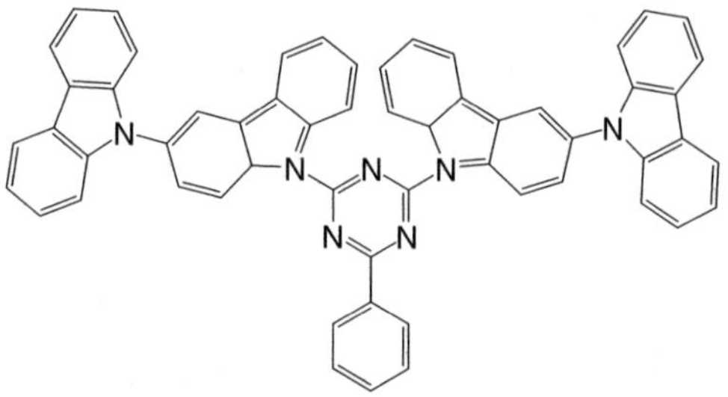

CC2TA

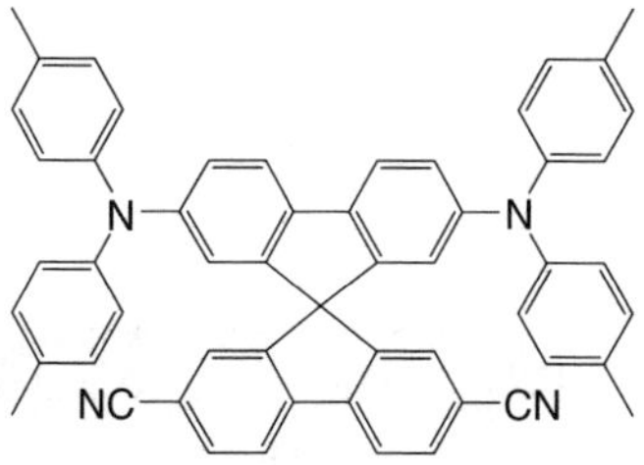

Spiro-CN

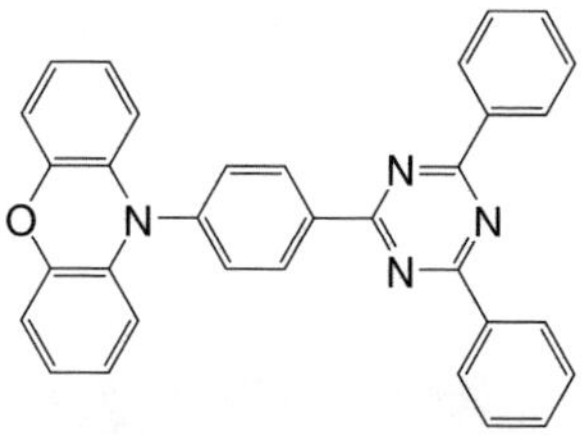

PXZ-TRZ

PBD

OXD-6

BMD

OXD-7

TAZ

PySPy

Figure 4.24   Some examples of classical electron transport materials

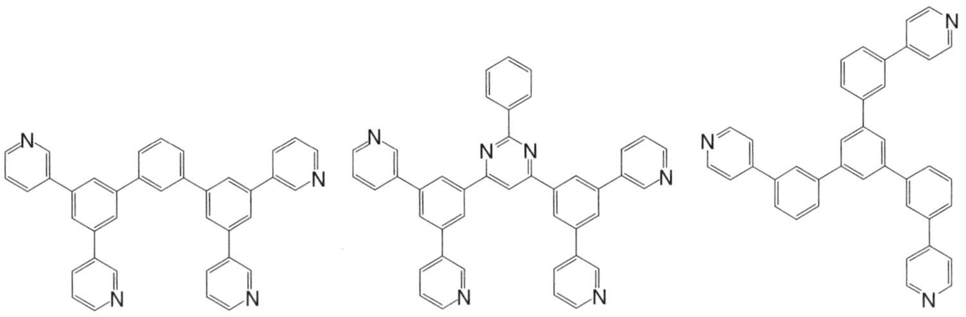

Figure 4.25 Recently developed electron transporting materials [57]

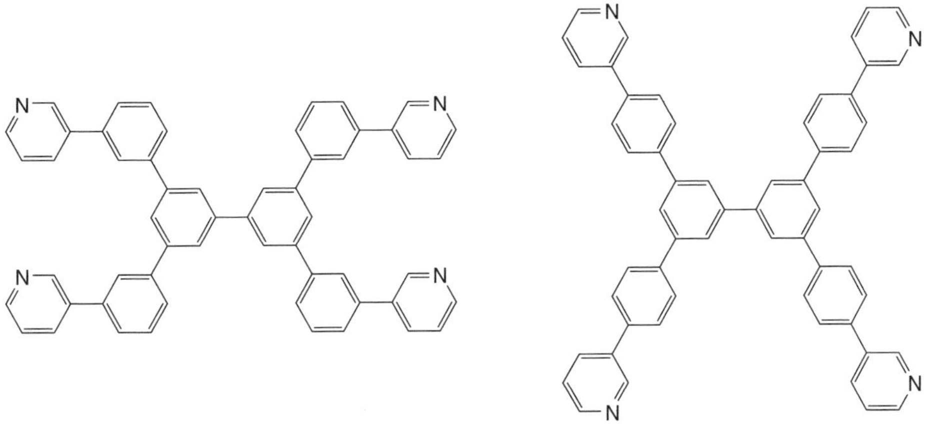

Figure 4.26 Recently developed electron transporting materials [57]

Recently, Li et al. of OITDA (Optoelectronics Industry and Technology Development Association) and Yamagata University (Japan) synthesized novel phenanthroline derivatives (Phens) as electron transport materials [55]. They reported that the Phens reduce the turn-on voltage, inducing higher efficiency than Alq$_3$.

For phosphorescent OLEDs, electron transport materials are required to have several properties such as high electron mobility, good hole-blocking ability, and sufficient triplet energy (E$_T$) for exciton blocking. Based on these requirements, novel electron transport materials have been synthesized [56, 57]. The examples are shown in Figs 4.25 and 4.26.

4.3.7 Electron Injection Materials and Cathodes

The significant role of the cathode is to inject electrons into an organic material which is adjacent with cathode. For this purpose, the work function of the cathode should be low. However, typical cathode metal such as Al, Ag, and ITO do not have a high work function. Therefore, in order to inject electrons into the adjacent organic layer effectively, an additional electron injection layer (EIL) is required. Indeed, EILs are very important in OLED devices because they play a role to reduce operating voltage significantly, by reducing the injection barrier between a cathode metal and an adjacent organic layer, which is usually an electron transport layer.

In actual OLED devices, a layer combining an electron injection layer and the cathode is often used, and this is often just called the cathode.

Several types of electron injection materials for EIL have been investigated and developed, one typical type of electron injection material being an inorganic electron injection layer. Typical examples are Mg:Ag [58], LiF [59–61], AlLi [62], and CsF [60], for small molecule OLEDs and Ca, Ba, etc. [63–67] for polymer OLEDs.

Wakimoto et al. of Pioneer (Japan) investigated the influence of various alkaline metal compounds on OLED performance [62]. They found that OLED devices with alkaline metal compounds showed reduced driving voltage and enhanced efficiency.

Jabbour et al. of Optical Sciences Center (USA) reported on the effects of several cathode materials on OLED characteristics [60]. According their report, although an Al cathode showed high driving voltage and very low efficiency, Mg, LiF/Al, Al-LiF composite, and Al-CsF composite showed reduced driving voltages and drastically improved efficiencies.

Application of an ultra-thin LiF/Al bi-layer in organic surface-emitting diodes was reported by Hung et al. [59]. Chen et al. reported that a stacked cathode layers composed of LiF(0.5 nm)/Al(1 nm)/Ag(20 nm) gave low sheet resistance (~1 Ω/sq), relatively low optical absorption, and good electron injection [68]. Okamoto et al. at Sharp successfully fabricated top emitting OLEDs with high efficiency, using micro-cavity effect induced by LiF(0.5 nm)/Al(1 nm)/Ag(20 nm) cathode [69].

The second type of electron injection material is ultra-thin metal such as Mg or Li at the organic-cathode interface. Kido et al. inserted an Mg (50 nm) or Li(1 nm) layer between an Ag cathode and an emitting Alq_3 layer, obtaining drastically enhanced luminance [70].

The third type is metal-doped organic layers [71, 72]. In 1998, Kido et al. in Yamagata University (Japan) reported on a metal-doped Alq_3 layer inserted between an Al cathode and non-doped Alq_3 layer [71]. Dopant metals are highly reactive metals such as Li, Sr, and Sm. A device with an Li-doped Alq_3 layer showed high luminance of over 30,000 cd/m^2, while a device without the metal-doped Alq_3 layer exhibited only 3400 cd/m^2. They suggested that the Li doping to the Alq_3 layer generates the radical anions of Alq_3 serving as intrinsic electron carriers, resulting in low barrier height for electron injection and high electron conductivity of the Li-doped Alq_3 layer.

The fourth type is electron injection layers composed of metal complexes such as 8-quinolinolato lithium (Liq, Fig. 4.27) [57, 73–76]. These metal complexes can be evaporated at relatively low temperatures of 200–300 °C and are easy to handle under ambient conditions. Endo et al. reported two types of methods for applying metal complexes to OLED devices [73]. The one is a simple insertion of a metal complex layer between a cathode metal and an organic layer. The other is an insertion of a layer with an organic material doped with a metal complex. Recently, Kido et al. in Yamagata University reported on lithium phenolate derivatives such as LiBPP, LiPP, and LiQP, as shown in Fig. 4.27 [57, 76].

4.3.8　Charge-Carrier and Exciton Blocking Materials

Charge-carrier and exciton blocking materials often play an important role in improving efficiency, lifetime, etc. In particular, it is well known that such charge-carrier and exciton blocking materials are very effective in phosphorescent OLEDs.

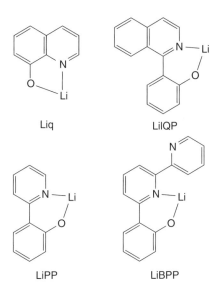

Figure 4.27 Examples of metal complex materials for EIL [57, 76]

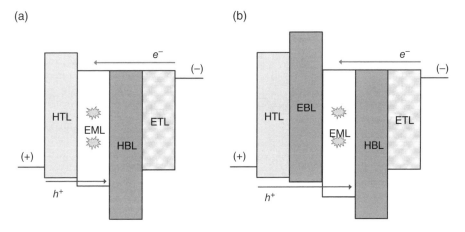

Figure 4.28 The role of charge-carrier and exciton blocking materials. Excitons are indicated as starburst patterns

The schematic energy diagrams for explaining the role of charge-carrier and exciton blocking materials are shown in Fig. 4.28.

Figure 4.28(a) illustrates the insertion of a hole and exciton blocking layer (HBL) between the electron transport layer (ETL) and the emission layer (EML). The HBL prevents hole and exciton leakage into the ETL, drastically improving the efficiency of OLED devices. The HOMO level of the HBL should be deeper than that of the EML.

Figure 4.28(b) illustrates the insertion of an electron and exciton blocking layer (EBL) between the hole transport layer (HTL) and the emission layer (EML). The EBL prevents electron and exciton leakage into the HTL, drastically improving the efficiency of OLED devices by balancing carrier injection into the EML. The LUMO level of the EBL should be higher than that of the EML. In addition, it is also important for the EBL to have a higher energy triplet level to prevent loss of excitons into the non-emissive adjacent HTL in phosphorescent OLED devices.

By using such charge-carrier and exciton blocking materials, carrier recombination and exciton confinement within the emission layer are efficiently forced.

Typical materials of HBL are shown in Fig. 4.29, one of which is the well-known material BAlq, aluminum(III)bis(2-methyl-8-quinolinato)4-phenylphenolate [77–79]. Kwong et al. reported that drastically improved lifetime was obtained in phosphorescent OLEDs by using a BAlq layer as hole and exciton blocking layer.

The other famous one, shown in Fig. 4.29 is 2,9-dimethyl-4,7-diphenyl-1,10-phenanthroline (bathocuproine BCP) [79–82]. The HOMO and LUMO of BCP are 6.5 eV and 3.2 eV, respectively [79]. Such a deep HOMO can effectively prevent hole transport.

Electron blocking materials are not often used in comparison with hole blocking materials because hole mobility is higher than electron mobility in many OLED devices. However, in some cases, electron blocking materials are useful and are also often required to have a high triplet energy level for preventing loss of excitons into the non-emissive adjacent HTL.

Typical materials of EBL are Irppz and $ppz_2Ir(dpm)$ [79], as shown in Fig. 4.30.

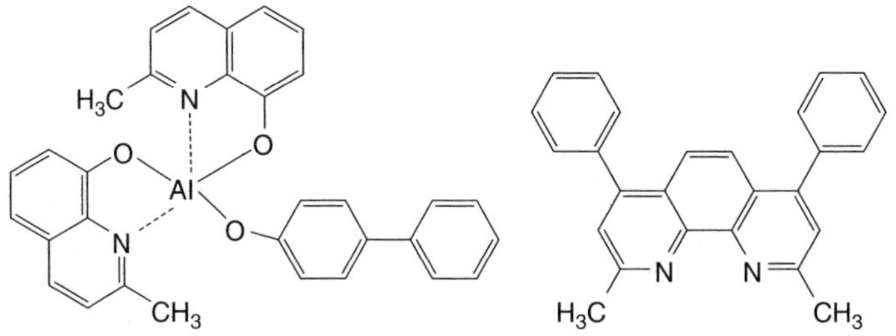

BAlq   BCP

Figure 4.29 Typical materials for HBL

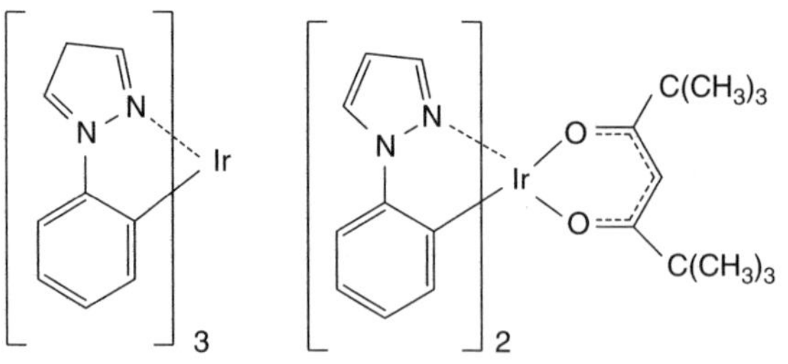

Irppz $ppz_2Ir(dpm)$

Figure 4.30 Typical materials for EBL

4.3.9 N-Dope and P-Dope Materials

It has been reported that p-i-n OLED devices are effective for obtaining high efficiency and low operating voltage [23, 83–86]. Such p-i-n OLED devices have structures in which an emission layer is sandwiched in between p- and n-type doped wide-band-gap carrier transport layers and appropriate charge blocking layers.

As is well known, conventional LEDs with inorganic semiconductors use heavily n- and p-doped electron and hole transport layers, leading to efficient tunneling injection and flat-band conditions under operation. Such device concepts can be applied to organic LEDs.

An example of a p-i-n OLED devices is shown in Fig. 4.31. The device has a p-doped hole injection and transport layer (HTL) and a n-doped electron transport layer (ETL). Due to the increased conductivity of organic semiconductor layers by doping with either electron donors for electron transport materials or electron acceptors for hole transport materials, the voltage drop across these layers can be significantly reduced. Such p-i-n type device structures guarantee an efficient carrier injection from both electrodes into the doped transport layers and low ohmic losses in these highly conductive layers.

Zhou et al. of Technische Universität Dresden (Germany) reported a p-doped hole injection layer doped with F_4-TCNQ [23], as is described in Section 4.3.1.

Fujihira and Ganzorig at the Tokyo Institute of Technology (Japan) reported that the p-doped hole transport TPD layer doped with oxidizing reagents such as iodine, $FeCl_3$, and $SbCl_5$, reduce the turn-on voltage, reducing the hole injection barrier [84]. In the ITO/TPD/Alq_3/Al device, the turn-on voltages when doped with these oxidizing reagents were reported to be lower than 10 V, while non-doped devices were reported to show higher turn-on voltage than 15 V.

Ikeda et al. of Semiconductor Energy Laboratory reported about molybdenum oxide (MoOx)-doped hole injection layers [85].

Su et al. of I-Shou University (Taiwan) also reported on molybdenum oxide (MoOx)-doped hole injection layers [86]. They doped MoOx to 4,4′,4,-tris[2-naphthyl(phenyl)amino] triphenylamine (2-TNATA), obtaining a p-doped hole injection layer (HIL). Their OLED

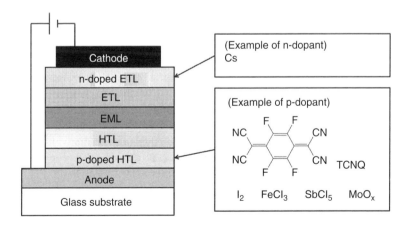

Figure 4.31 An example p-i-n OLED device [23]

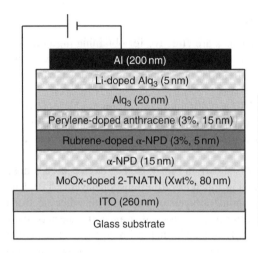

X (wt%)	Power efficiency (lm/W)
0	2.95
5	3.55
10	4.32
15	3.49
20	3.05

Figure 4.32   OLED device structures utilizing the p-doped HIL and their typical performances. [86]

device structures utilized the p-doping HIL, and the typical performances are shown in Fig. 4.32. The results indicate that the MoOx doping improves the power efficiency and reduces the driving voltage. In addition, they also reported that the lifetime was drastically improved.

On the other hand, as n-doped layers, co-evaporation of BPhen with pure Cs [84] has been reported.

4.4   Solution Materials

Solution processes for fabricating OLED devices have attracted much attention because they can provide breakthrough for low cost OLED production technologies. Solution processes require solution type OLED materials, which are classified into three types: polymers, dendrimers, and soluble small molecules.

4.4.1   Polymer Materials

Polymer OLEDs were first reported by Burroughes et al. of the Cambridge group [87, 88]. They used poly(p-phenylene vinylene), PPV, as shown in Fig. 4.33. PPV is a conjugated organic semiconductor, in which the π molecular orbitals are delocalized along the polymer chain. Burroughes et al. wrote that the injection of an electron and a hole on the conjugated chain can lead to a self-localized excited state which can then decay radiatively, suggesting the possibility of using these materials in electroluminescent devices. For preparing PPV, they synthesized a solution-processable precursor-polymer (II) as shown in Fig. 4.33. The precursor-polymer II was spin-coated on substrates with an indium oxide, followed by thermal conversion (typically >250 °C, in *vacuo*, for 10 hours) to the homogeneous, dense, and uniform film of PPV (I) with a typical thickness of 100 nm. As a cathode, Al was deposited on the polymer. The device structure is also shown in Fig. 4.33.

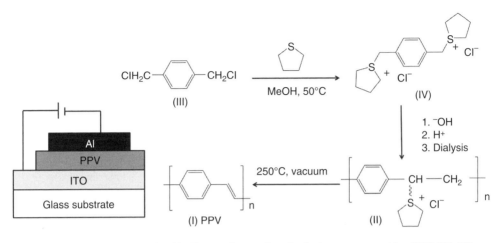

Figure 4.33 PPV synthesized by Burroughes et al. and a device structure with a PPV. [87, 88]

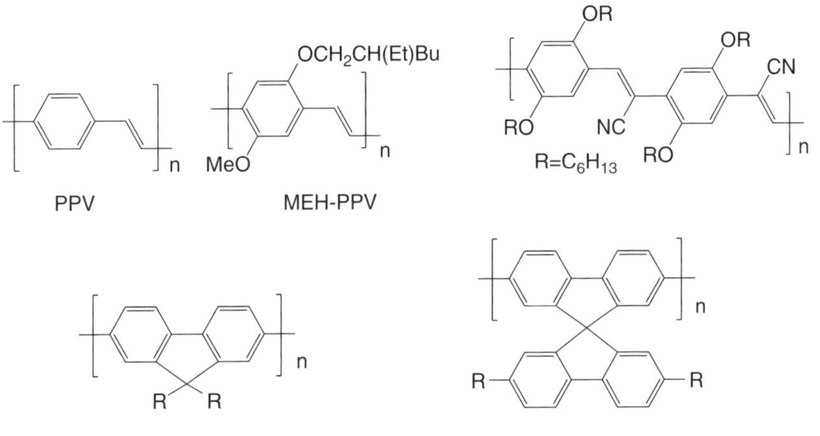

Figure 4.34 Examples of conjugated polymers utilized in polymer OLED devices

After the report by Burroughes et al., active research and development started. While the organic layer in the device reported by Burroughes et al. is only one layer, multi-layer structures and various materials have been developed.

4.4.1.1 Fluorescent Emitting Polymers

Examples of conjugated polymers utilized in polymer OLED devices are summarized in Fig. 4.34.

Poly(p-phenylene vinylene), PPV was used in the first report of electroluminescence from conjugated polymers [87, 88]. PPV has an energy gap between the π and π^* states of about 2.5 eV, producing yellow/green luminescence. Several PPV derivatives shown in Fig. 4.34 have been synthesized and applied to polymer OLED devices. One of the famous PPV

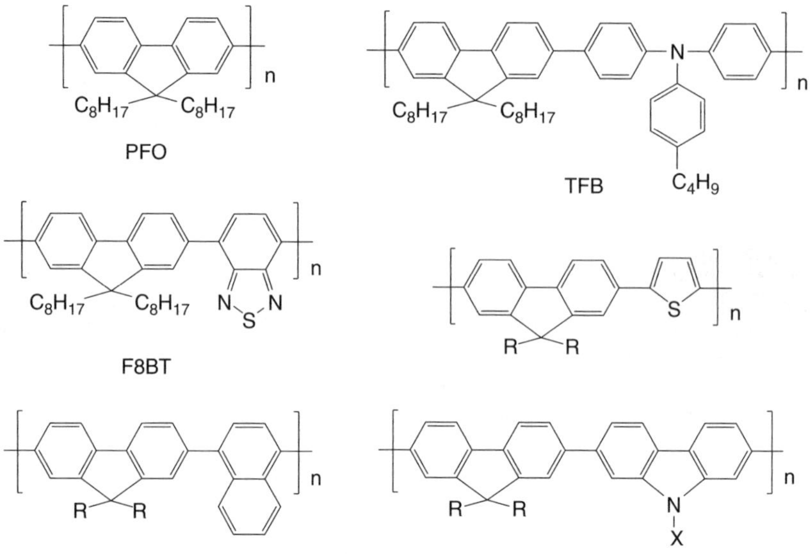

Figure 4.35   Examples of copolymers with polyfluorene. [93]

derivatives is MEH-PPV, poly(2-methoxy,5-(2′-ethyl-hexoxy)-1,4-phenylene-vinylene), which generates orange-red color and has been used in many studies, because MEH-PPV is soluble in organic solvents [89]. MEH-PPV has a π-π* electronic energy gap of about 2.2 eV, which is lower than that of PPV. Copolymers have been widely developed and applied to polymer OLED devices because they allow color tuning and can show improved luminescence [90].

Poly(dialkylfluorene)s [91–93] shows blue emission and high luminescence. Various copolymers with poly(dialkylfluorene) have been synthesized and applied to OLED devices. Some copolymers of polyfluorene realized green and red colors. Examples are shown in Fig. 4.35.

4.4.1.2   Hole Injection Materials

As well as small molecular OLED devices, the hole injection property is important in polymer OLED devices. For reducing the hole injection barrier, hole injection materials are often deposited on the anode (normally ITO).

Yang and Heeger et al. of Uniax Corporation (USA) reported that charge carrier injection from an ITO anode is improved by inserting a hole injecting layer which consists of blends of polyaniline (PANI) in low molecular weight polyester resin [94].

Cao et al. of Heeger's group applied polyethylene dioxythiophene-polystyrene sulfonate (PEDOT:PSS) thin films to polymer light-emitting diodes with poly(2-methoxy,5-(2′-ethyl-hexyloxy)-1,4-phenylene vinylene) (MEH-PPV) [95]. The OLED devices with an ITO/PEDOT:PSS/MEH-PPV/Ca structure were reported to show improved efficiency and lifetime.

The molecular structure of PEDOT:PSS is shown in Fig. 4.36. PEDOT:PSS has been widely used not only for polymer type OLEDs but also for combination with evaporated organic layers [96–101].

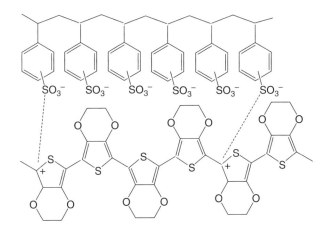

Figure 4.36   Molecular structure of PEDOT:PSS

(a)   (b)

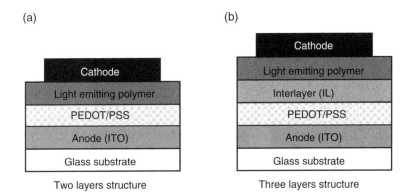

Two layers structure   Three layers structure

Figure 4.37   Two typical device structures of polymer OLED devices with PEDOT:PSS. (a) Two-layer structure without an interlayer. (b) Three layer structure with an interlayer

4.4.1.3   Degradation of PEDOT:PSS and Interlayer

Although PEDOT:PSS is an useful solution material with hole injection and hole transporting abilities, one of its serious problems has been the negative influence on OLED lifetime under driving.

As the counter technology for utilizing PEDOT:PSS, an insertion of an interlayer between PEDOT:PSS and an emitting layer has been proposed. Figure 4.37 shows typical two device structures of polymer OLEDs with PEDOT:PSS. Figure 4.37(a) shows the two-layer structure without an interlayer and Fig. 4.37(b) shows the three-layer structure with an interlayer. The effect of the interlayer has been reported as the prevention of exciton quenching [102] and/or electron blocking [103–107].

Based on the electron blocking model, schematic typical energy diagrams of these two device structures, with or without an interlayer, are shown in Fig. 4.38. When there is no inter-layer, as shown in Fig. 4.38(a), electron can be transported to PEDOT:PSS easily. On the other hand, if the interlayer prevents the electron transport to PEDOT:PSS, electrons do not tend to

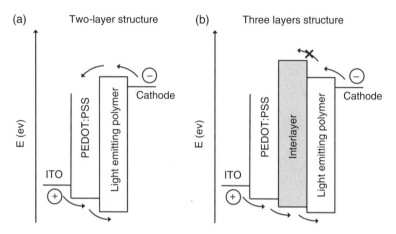

Figure 4.38 Typical schematic energy diagrams of two device structures of polymer OLED devices with or without an interlayer. (a) Two-layer structure without an interlayer (b) Three-layer structure with an interlayer

be transported to PEDOT:PSS, as shown in Fig. 4.38(b). The idea of the electron blocking interlayer is closely related to the degradation mechanism of PEDOT:PSS.

For investigating the degradation mechanism of PEDOT:PSS, Sakamoto et al. in Furukawa's group of Waseda University investigated the degradation of PEDOT:PSS in polymer type OLED devices by using Raman spectral changes [108]. Their device structure is Glass/ITO(~100 nm)/PEDOT:PSS(~30 nm)/F8BT-PF8(~90 nm)/Li-Al. They reported finding an increase of the PEDOT band under driving, and then concluded that this phenomenon can be attributed to the reduction (dedoping) of the PEDOT chains. They also explained that a proportion of the electrons injected into the F8BT-PF8 blende layer may escape recombination and reach the PEDOT:PSS layer and then the PEDOT chains may be reduced (dedoped). They suggested that this phenomenon is most likely one of the intrinsic factors in the degradation of polymer OLED devices with PEDOT:PSS. Indeed, there are several reports on the improvement in efficiency and/or lifetime by the insertion of interlayer between PEDOT:PSS and light emitting polymer.

Morgado et al. reported on the electron blocking layer in polymer OLEDs [103]. They inserted a thin film of poly(p-phenylene vinylene), PPV, between a hole-injection layer of PEDOT:PSS and the polyfluorenes emissive layer. Figure 4.39 shows the schematic energy diagram of their device with ITO/PEDOT:PSS/PPV/EML/Ca, where EML is PFO with 5 wt% of green emitting F8BT. They reported that the efficiency increases from 2.1 to 4.1 cd/A by inserting a PPV layer. At the F8BT/PPV interface, there is a barrier of about 0.4 eV for electron injection into PPV due to the significant mismatch of the respective LUMO levels. They proposed that such an improvement is mainly due to the electron-blocking effect of the PPV layer, leading to improved charge carrier balance within the emissive layer.

Conway et al. of Cambridge Display Technology (CDT) investigated the role of the interlayer by using an electron-only current device with an interlayer with different electron mobility [104]. They observed that the EQE and lifetime increase with decreasing the electron current of interlayer, concluding that the transport properties of the interlayer and light emitting polymer are of key importance, and the interlayer exciton blocking properties are not as critical.

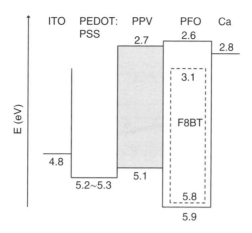

Figure 4.39   The schematic energy diagram of the device with ITO/PEDOT:PSS/PPV/EML/Ca, where EML is PFO with 5\wt% of green emitting F8BT [103]

Table 4.1   Effects of an insertion of interlayer [106, 107]

		Efficiency (cd/A)		Half lifetime (h)	
			Luminance		Initial luminance
Red devices	Two-layer (without IL)	2.3	450 cd/m²	98 h	3000 cd/m²
	Three-layer (with IL)	3.0		569 h	
Green devices	Two-layer (without IL)	13.0	900 cd/m²	130 h	6000 cd/m²
	Three-layer (with IL)	14.1		378 h	
Blue devices	Two-layer (without IL)	5.7	150 cd/m²	15 h	1000 cd/m²
	Three-layer (with IL)	11.2		464 h	

The device structure: ITO/PEDOT:PS(65 nm)/LEP(80 nm)/Ba/Al.
ITO/PEDOT:PS(65 nm)/IL(20 nm)/LEP(80 nm)/Ba/Al.

Shirasaki et al. of Casio (Japan) also reported applying an interlayer as an electron blocking interlayer to their polymer AM-OLED display [105].

Fujita et al. in Sharp (Japan) obtained improvement in efficiency and lifetime of red, green, and blue polymer OLED devices with the interlayer [106]. While the improvement in efficiencies of red and green polymer OLED devices are not so large, the efficiency improvement of the blue OLED is drastic, being twice. In addition, as shown in Table 4.1, it was found that the lifetimes were drastically improved in red, green, and blue OLEDs. One of the examples of the improvement in lifetime curves is shown in Fig. 4.40.

Hatanaka et al. at Sharp (Japan) investigated the role of the interlayer, using single carrier devices and bipolar devices [107]. The device structures of their single carrier devices are shown in Figs 4.41 and 4.42. In red, green, and blue polymer OLED devices, the same interlayer material is used with the same thickness of 20 nm. The V–I curves in electron-only devices (EODs) are shown in Fig. 4.41. In the V–I curves in EODs, the currents are remarkably reduced by the insertion of the interlayer. Figure 4.41 clearly shows that the interlayer effectively blocks electrons. However, this phenomenon can not be explained by the difference

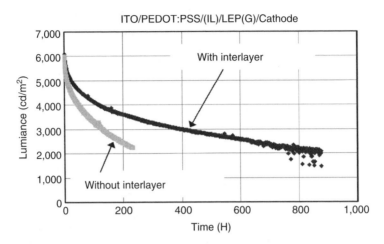

Figure 4.40 An example of lifetime improvement by an insertion of an interlayer

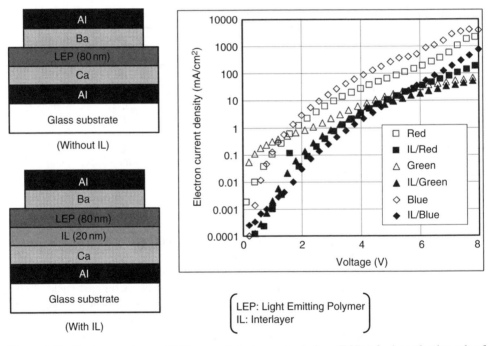

Figure 4.41 Device structure and I–V cures of electron-only devices (EODs) for investigating role of interlayers

in LUMO levels. Figure 4.43 shows the energy diagram. If the electron blocking occurs due to the gap in LUMO levels between the interlayer and the emitting polymer, the electron blocking effect should be largest in the red OLED and smallest in the blue OLED. However, Fig. 4.41 shows that the electron blocking in the blue OLED is largest. Therefore, it is concluded that the electron blocking effect due to the interlayer is not attributed to the difference

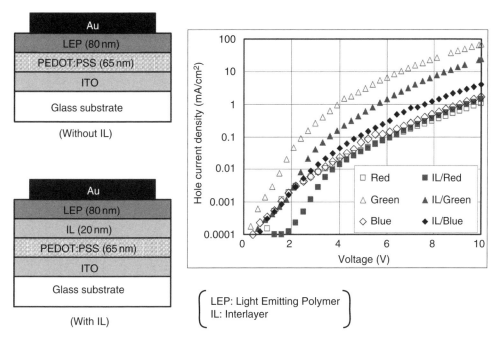

Figure 4.42 Device structure and I–V cures of hole-only devices (HODs) for investigating role of interlayers

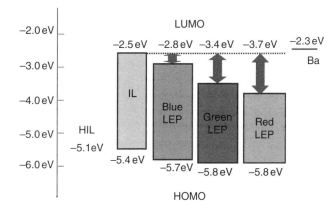

Figure 4.43 The energy diagram of several polymer materials

in LUMO levels between the interlayer and the emitting polymer. On the other hand, it should be noticed that the red, green, and blue OLED devices with the interlayer show small differences in the current densities, while those without the interlayer show large differences. This means that low electron mobility of the interlayer gives rise to electron blocking, giving the same level of current density in three color OLED devices. They also investigated hole only devices (HODs), as shown in Fig. 4.42, but found no clear relationship with the improving effect of the interlayer. These investigations using single carrier devices indicate that the interlayer plays a role in electron blocking due to the low electron mobility, inducing the drastic improvement in efficiency and lifetime.

4.4.1.4 Phosphorescent Polymer Materials

The solution type phosphorescent polymer materials have also been studied, the first generation of being a mixture of a host polymer with a small molecule phosphorescent dopant.

Guo et al. in University of California at Los Angeles and University of Southern California (USA) doped small molecular platinum(II)-2,8,12,17-tetraethyl-3,7,13,18-tetramethylporphyrin (PtOX) to a host polymer [109]. The device structure is a bi-layer structure consisting of a hole transporting layer and an electron transporting layer. The hole transporting layer consists of poly(vinylcarbazole). The electron transporting layer consists of poly(9,9-bis(octyl)-fluorene-2,7-diyl) (BOc-PF) doped with PtOX, which is a phosphorescent emitting dopant. The external efficiency of the OLED device was reported to be enhanced from 1% to 2.3% when doped with PtOX.

Lee et al. in Kwangju Institute of Science and Technology (Korea) reported a phosphorescent OLEDs with tris(2-phenylpyridine) iridium [Ir(ppy)$_3$] as a triplet emissive dopant in poly(vinylcarbazole) (PVK) host [110]. Their device structure is shown in Fig. 4.44. The device with 8% doping concentration of [Ir(ppy)$_3$] in PVK showed an EQE of 1.9% and the peak luminance of 2500 cd/m^2.

Lamansky and coworkers at University of Southern California (USA) also reported phosphorescent OLEDs consisting of poly(N-vinylcarbazole) (PVK)-based single-layer doped with small molecular phosphorescent dyes [111].

Lane et al. investigated polymer OLEDs with a luminescent polymer host, poly(9,9-dioctyl-fluorene) (PFO) doped with a red phosphorescent dye, 2,3,7,8,12,13,17,18-octaethyl-21H,23H-porphyrin platinum(II) (PtOEP) [112]. They obtained a maximum EQE of 3.5% at a PtOEP doping concentration of 4 wt%.

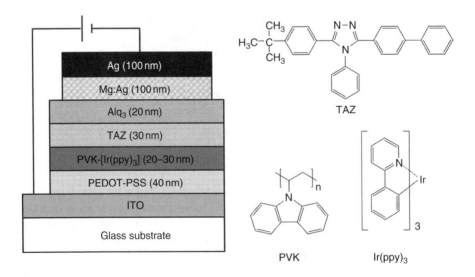

Figure 4.44 Device structure of OLEDs with a host polymer and a phosphorescent dopant [110]

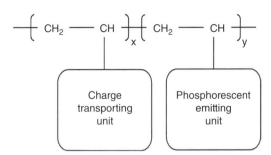

Figure 4.45 Schematic illustration of second generation phosphorescent polymers

Vaeth and Tang of Eastman Kodak (USA) fabricated phosphorescent OLEDs consisting of a poly(vinyl carbazole) host doped with fac tris(2-phenylpyridine) iridium (Ir-ppy) [113]. They achieved a luminance efficiency of 30 cd/A and an EQE of 8.5% in the optimized condition with a second dopant of 2-(4-biphenylyl)-5-(4-tert-butylphenyl)-1,3,4-oxadiazole.

Hino et al. in Osaka University (Japan) reported on phosphorescent OLED devices with a phosphorescent emitter, $Ir(ppy)_3$ doped into a solution-processable low molecular material [114]. As a host material, they used methoxy-substituted 1,3,5-tris[4-(diphenylamino)phenyl] benzene (TDAPB), which was dissolved 1,2-dichloroethane. They obtained a peak EQE of 8.2% and a current efficiency of 29 cd/A.

The second generation of solution-type phosphorescent materials are emissive polymers involving a small amount of phosphorescent units in the side group of the polymers. Such materials are schematically illustrated in Fig. 4.45.

Lee and coworkers at Kwangju Institute of Science and Technology (Korea) synthesized a new polymer containing carbazole units and iridium complexes for fabricating phosphorescent OLED devices [115]. The content of the Ir complex is 7.8 wt% with respect to the carbazole unit in the polymer. Their device structure is ITO/PEDOT(40 nm)/Ir complex copolymer (30 nm)/TAZ(30 nm)/Alq_3(20 nm)/LiF(1 nm)/Al(180 nm). They reported that the device showed a maximum EQE of 4.4% at 36 cd/m^2, a maximum power efficiency of 5.0 lm/W at 6.4 V, and a peak luminance of 12,900 cd/m^2 at 24.2 V (360 mA/cm^2).

As a collaborative study between NHK Science and Technical Research Laboratories and Showa Denko (Japan), Tokito et al. also developed phosphorescent polymers involving a carbazole unit and an iridium-complex unit [116]. The molecular structures are shown in Fig. 4.46. They fabricated OLED devices with the structure of glass/ITO/PEDOT:PSS(30 nm)/ EML(85 nm)/Ca(10 nm)/Al(150 nm), where the emission layer (EML) involves a phosphorescent polymer doped with an electron transport material. High external quantum efficiencies of 5.5%, 9%, and 3.5% were obtained in red, green and blue OLEDs respectively.

Tokito et al. of NHK Science and Technical Research Laboratories improved the performance of phosphorescent OLED devices with a phosphorescent polymer by inserting aluminum(III) bis(2-methyl-8-quinolinato)4-phenylphenolato (BAlq) as hole blocking layer between an emissive polymer layer and a cathode [117]. They achieved external quantum efficiencies of 6.6%, 11%, and 6.9% in red, green and blue OLEDs, respectively.

In addition, they developed improved polymers. Suzuki et al. synthesized novel phosphorescent copolymers [118]. The copolymers have bis(2-phenylpyridine)iridium (acetylacetonate) [Ir(ppy)(2)(acac)], N,N′-diphenyl-NN′-bis(3-methylphenyl)-[1,1′-biphenyl]-4,4′-diamine

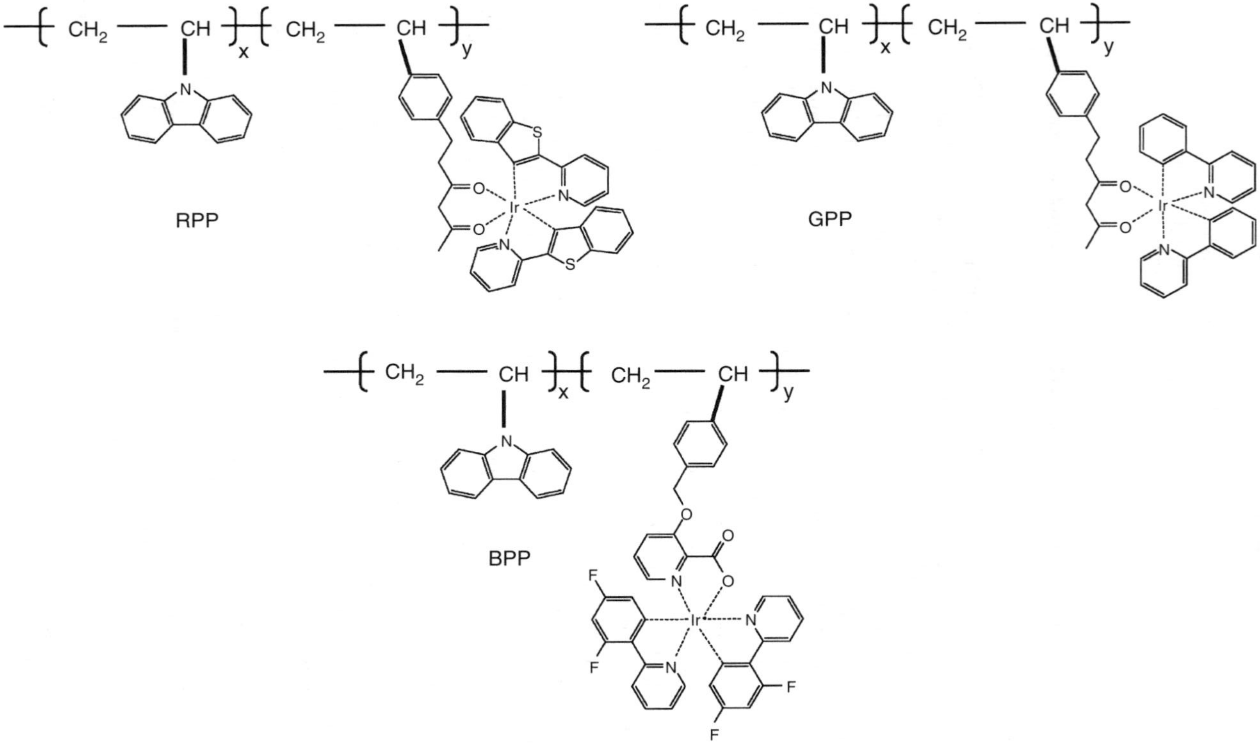

Figure 4.46 Molecular structures of phosphorescent polymer involving a carbazole unit and an iridium-complex unit developed by Tokito et al. [116]

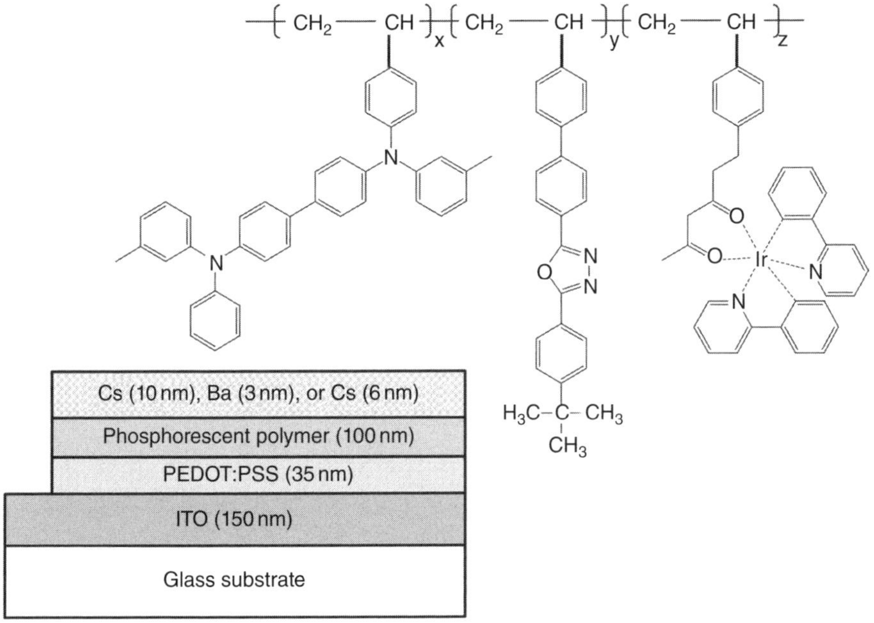

Figure 4.47 Device structure of OLEDs with a phosphorescent polymer reported by Suzuki et al. [118]

(TPD) and 2-(4-biphenyl)-5-(4-tert-butylphenyl)-1,3,4-oxadiazole (PBD) as a side group. Figure 4.47 shows the molecular structure of the developed polymer and the OLED device structure with the polymer. They investigated the influence of concentration ratio of three types of substituents and three low work function metals, Ca, Ba, and Cs, as the electron injection layer. In the best case, the combination of the materials with the ratio of TPD:PBD:Ir(ppy)2(acac) = 18:79:3 and Cs gives an EQE of 11.8% and a power efficiency of 38.6 lm/W.

4.4.2 Dendrimers

Dendrimers are also solution type materials, the concept of which is illustrated in Fig. 4.48. Dendrimers consist of a core, conjugated dendrons, and surface groups. The core at the center of the dendrimer molecule provides opto-electronic properties, which are essentially related to OLED performances. The dendrons can control charge mobility, etc. The surface groups on the outside of the molecule, on the other hand, affect the solubility and processability. Dendrimers have well-defined structure and can be purified by general purification techniques, while structural defects in polymers cannot be easily removed and those defects could result in poor device stability.

The first approaches of dendrimers for OLEDs were fluorescent dendrimers. Halim et al. of University of Durham and Dyson Perrins Laboratory (UK) reported conjugated dendrimers shown in Fig. 4.49 [119]. These materials were applied to simple OLED devices with the structure of glass/ITO/dendrimer/cathode, where the dendrimer is spin-coated and the metal cathode is evaporated. They reported that green, red and blue OLEDs were obtained by changing the molecular structure of dendrimers.

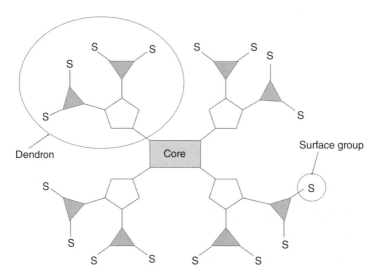

Figure 4.48   The dendrimer concept showing the core, conjugated dendrons, and surface group

Lupton et al. of University of St Andrews etc. (UK) reported charge transporting dendrimers [120] shown in Fig. 4.50.

The second approach with dendrimers involves phosphorescent dendrimers. The phosphorescent dendrimers generally consist of an emitting core with a heavy metal and surrounding dendrons. These dendrons play the roles of transporting charge carriers and suppressing intermolecular interactions between emitting iridium cores, which cause photoluminescent quenching in a neat film. In addition, the solubility of dendrimer metal complex is generally better than that of conventional metal complexes. One of the requirements for a dendron is to possess a higher triplet excited energy level than that of the core complex so as not to quench triplet emission from the core complex.

In 2001, Lupton et al. reported a phosphorescent dendrimer with Platinum (Pt) as a heavy metal at the center of the dendrimer [121].

Markham et al. at University of St Andrews (UK) reported phosphorescent dendrimers with iridium (Ir) as the heavy metal at the center of the dendrimer [122]. The molecular structures are shown in Fig. 4.51. They fabricated OLED devices with the dendrimer, as shown in Fig. 4.52. The emission layer consists of a TCTA and the dendrimer G1. On the dendrimer layer, an electron transporting layer of TPBI is deposited. They obtained a maximum efficiency of 40 lm/W (55 cd/A) at 4.5 V and 400 cd/m^2.

Sumitomo Chemical (Japan) and Cambridge Display Technology (CDT) (UK) have also reported solution type emitting materials containing a phosphorescent dendrimer and conjugated host polymer [123]. The schematic molecular structure is shown in Fig. 4.53. The dendrimers possess a core with an Ir complex exhibiting phosphorescent emission and a dendron with aromatic moieties having groups for solution-compatible groups, exhibiting good emission properties, good charge transport properties and good solubility. The host polymers are polyfluorene polymers having excellent charge carrier transport ability and solution processability. They doped a red dendrimer to the host polymer with blue fluorescent emission. The structure of OLED devices

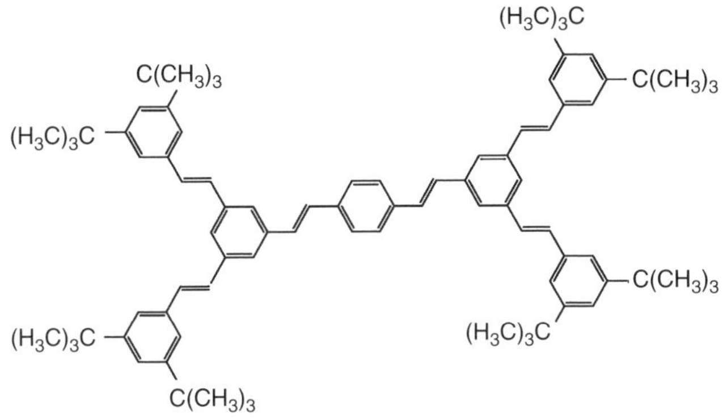

1-DSB

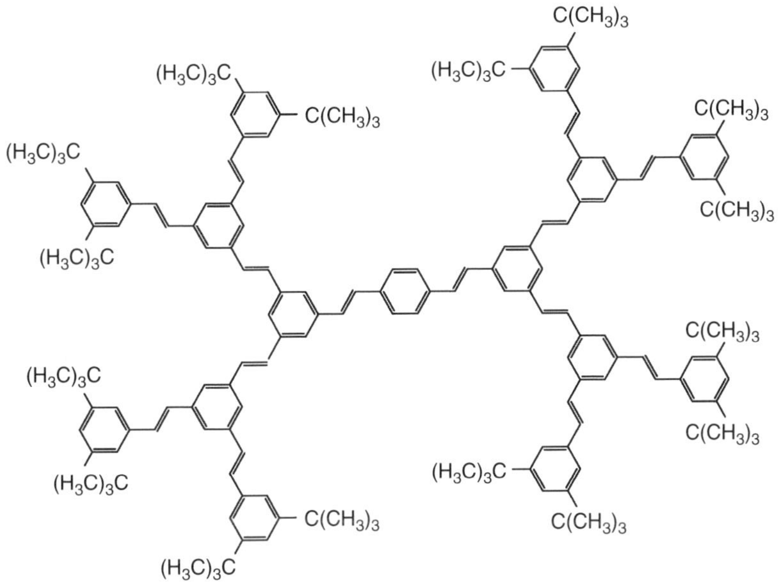

2-DSB

Figure 4.49 Fluorescent dendrimers reported by Halim et al. [119]

with the dendrimers is shown in Fig. 4.54. They obtained a current efficiency of 4.6 cd/A, CIE color ordinate of (0.66, 0.32), and half lifetime (T_{50}) of 5700 hours from the initial luminance of 400 cd/m².

Tsuzuki et al. at Tokito's group at NHK Japan Broadcasting (Japan) developed phosphorescent dendrimers that have a phosphorescent core and dendrons based on charge-transporting building blocks [124]. They synthesized first-generation and second-generation dendrimers consisting of a fac-tris(2-phenylpyridine)iridium [Ir(ppy)₃] core and hole-transporting

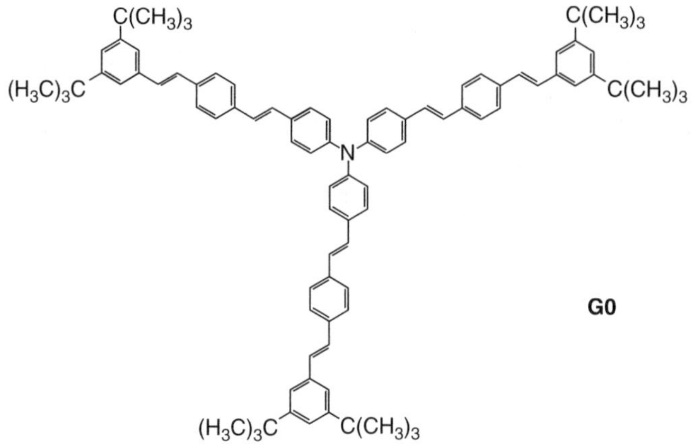

Figure 4.50   Charge transporting dendrimers reported by Lupton et al. [120]

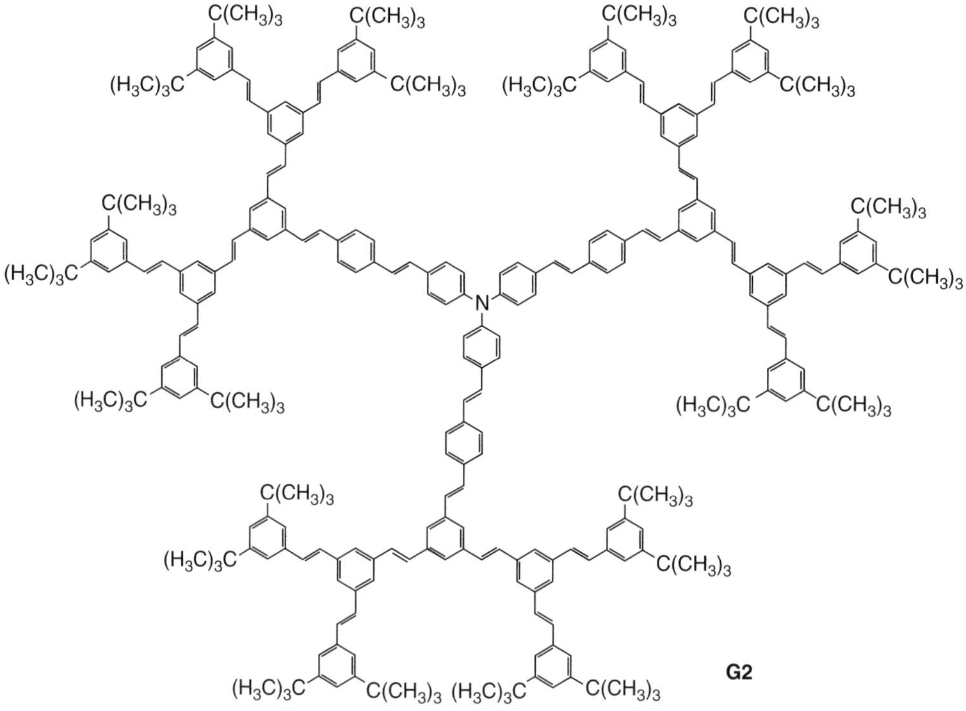

phenylcarbazole-based dendrons. The OLEDs using the film containing a mixture of the den-drimer and an electron-transporting material was reported to exhibit bright green or yellowish green emission with the maximum EQE of 7.6%.

Iguchi et al. at Yamagata University (Japan) synthesized solution-processable iridium complexes having bulky carbazole dendrons [125]. The synthesized materials are shown in Fig. 4.55. The ligand has a 4-pyridine-substituted dendron to improve electron injection from the cathode

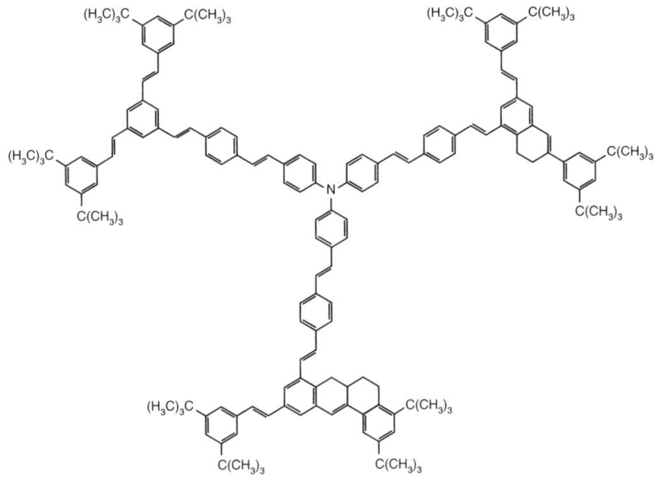

Figure 4.50 (*Continued*)

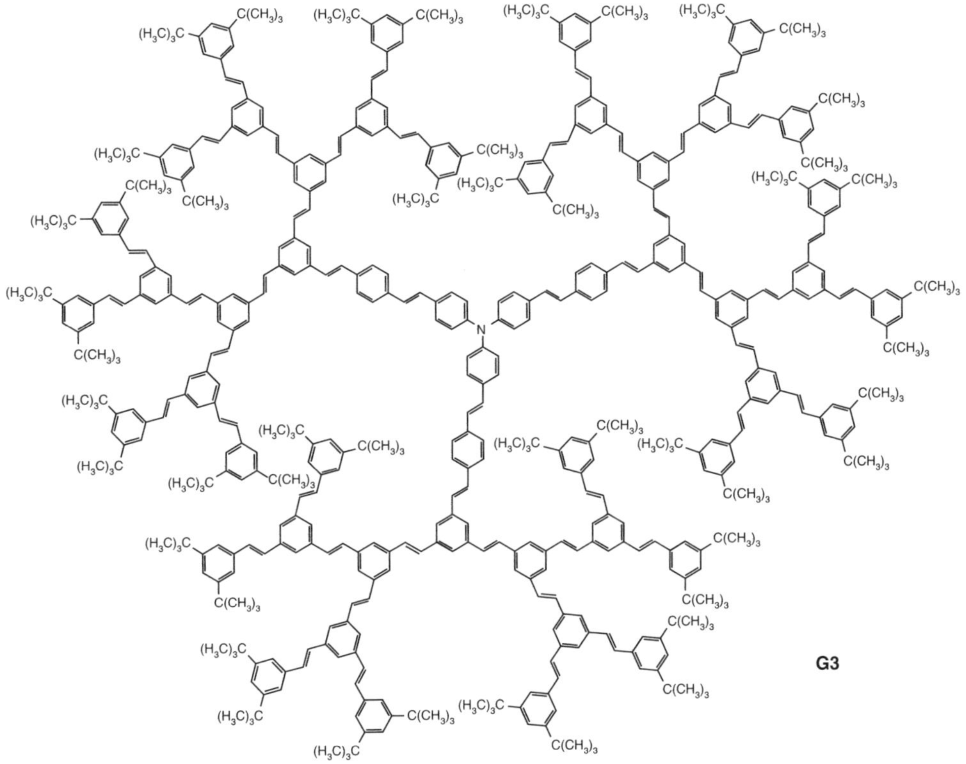

metal. Suppression of concentration-quenching and efficient exciton confinement by the bulky dendrons having higher triplet energy than that of Ir(ppy)$_3$ were reported. Photoluminescence quantum efficiencies (PLQEs) of (mCP)$_3$Ir and (mCP)$_2$(bpp)Ir in their diluted solutions were 91% and 84%, respectively. They showed high PLQEs of 49% for (mCP)$_3$Ir and 29% for (mCP)$_2$(bpp)

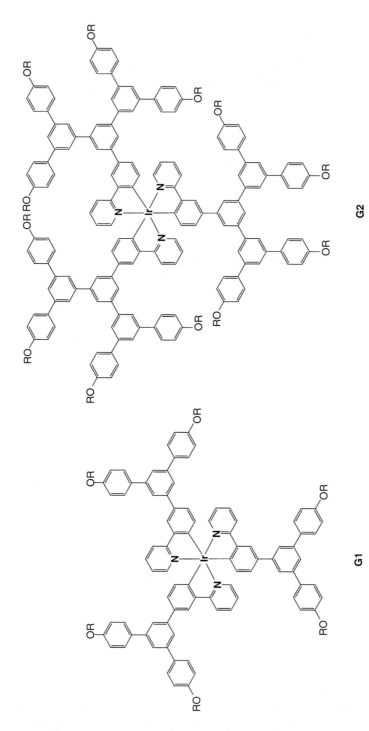

G1

G2

Figure 4.51 Molecular structure of phosphorescent dendrimers reported by Markham et al. [122]

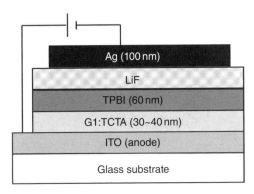

Figure 4.52   OLED device structure with a dendrimer reported by Markham et al. [122]

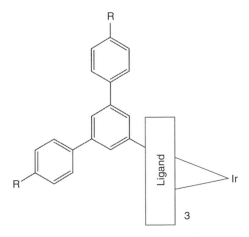

Figure 4.53   Molecular structures of dendrimers and host polymers reported by Pillow et al. [123]

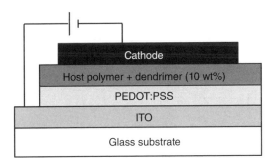

Figure 4.54   The OLED device with a dendrimer reported by Pillow et al. [123]

Ir even in a neat film. The triplet exciton energy levels of the dendronized ligand (2.8 eV) and the dendron (2.9 eV) are sufficiently higher than that of the core complex Ir(ppy)$_3$ (2.6 eV). When an electron-transporting and hole-blocking material was used, the EQEs of double-layer devices were dramatically improved to 8.3% for (mCP)$_3$Ir and 5.4% for (mCP)$_2$(bpp)Ir at 100 cd/m^2.

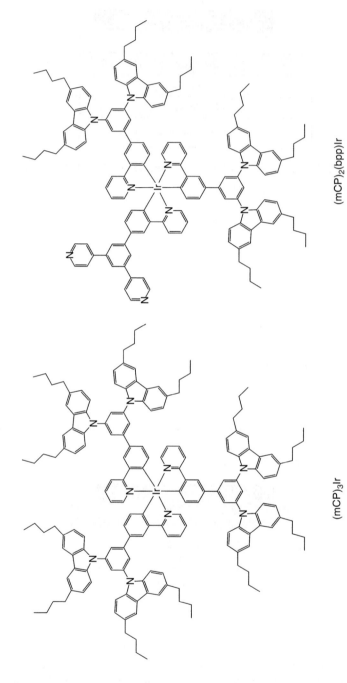

(mCP)₃Ir

(mCP)₂(bpp)Ir

Figure 4.55 Solution-processable iridium complexes having bulky carbazole dendrons reported by Iguchi et al. [125]

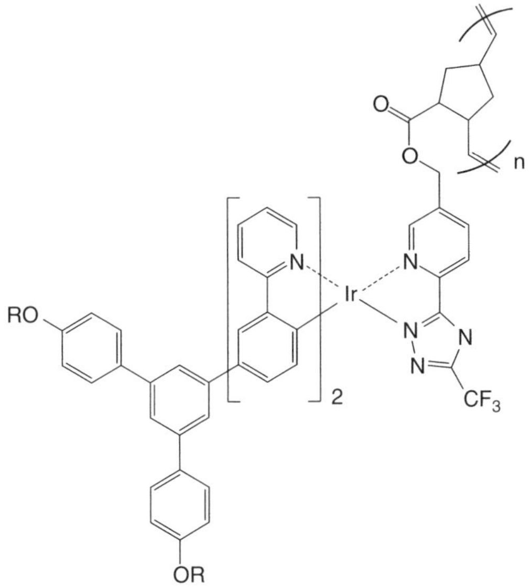

Figure 4.56 A phosphorescent green solution-processable iridium(III) complex containing poly(dendrimer) reported by Levell et al. [126]

Table 4.2 Device performance OLED devices with solution-based small OLED materials [127]

Color	Current efficiency (cd/A)	Voltage (V)	CIE (x, y)	LT_{50} (hr)
Deep red	14.0	4.6	(0.68, 0.32)	240,000
Red	20.3	4.1	(0.65, 0.35)	200,000
Green	68.3	3.5	(0.32, 0.63)	125,000
Blue	6.2	3.9	(0.14, 0.14)	24,000
Deep blue	3.2	5.1	(0.14, 0.08)	9,000

All data at 1000 cd/m^2 and 20 °C.

Levell et al. of University of St Andrews (UK), University of Oxford (UK) and The University of Queensland (Australia) reported a phosphorescent green solution-processable iridium(III) complex containing poly(dendrimer). They achieved an EQE of 5.1% and current efficiency of 16.4 cd/A in an OLED device with a poly(dendrimer) shown in Fig. 4.56 [126].

4.4.3 Small Molecules

The third approach for solution-processed OLED materials is soluble small molecule OLED materials. Herron and Gao of DuPont Displays (USA) reported on their solution-based small OLED materials [127]. Table 4.2 shows the performance of the OLED devices fabricated by using spin-coating of their solution-based small OLED materials. The data obtained, assume "common layers architecture", as shown in Fig. 4.57, in order to manufacture in a cost-effective manner. Therefore, the thickness of HIL, HTL, and ETL is common across all three colors. It is understood that Table 4.2 shows good device performance in all three colors.

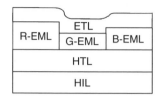

Figure 4.57   Illustration of the common layers architecture used in full-color displays

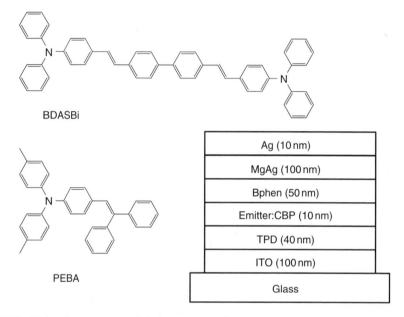

Figure 4.58   Molecular structures and device structure of the materials reported by Frischeisen et al. [129]

4.5   Molecular Orientation of Organic Materials

For a long time, it has been believed that organic thin films for OLED devices should be amorphous because the presence of randomly rough interfaces, such as polycrystalline structure, induces an undesirable current barrier, leak current between anode and cathode, and undesirable light scattering. In addition, if any special techniques are not applied, it has also been supposed that films deposited by vacuum deposition have amorphous molecular orientation.

However, it has recently been reported that molecular orientation of vacuum deposited or wet coated films has been obtained in some cases. Molecular orientations of organic thin film are interesting not only from a scientific point of view but also from a practical point of view, because molecular orientations can induce improvements in carrier mobility, carrier injection, and light extraction efficiency.

Yokoyama et al. at Kyushu University (Japan) discovered that organic materials in evaporated organic layers can have anisotropic molecular orientation, investigating molecular orientation of organic films by using variable angle spectroscopic ellipsometry (VASR) [128].

   Using the effect of molecular orientation of emitting materials, Frischeisen et al. at Kyushu University (Japan) reported a significant enhancement of light out-coupling efficiency [129]. They used two materials, PEBA and BDASBi, that are very similar except for their orientation. Figure 4.58 shows the molecular structure and the device structure to which these materials are applied. They evaluated and determined dipole orientation in guest–host systems and found that PEBA doped into CBP shows no preferred molecular orientation and the dipole orientation is completely random (i.e. on average 2/3 of the dipoles are horizontally oriented and 1/3 vertically) but the sample containing BDASBi shows a clear tendency towards horizontal orientation (random and horizontal dipole orientations with 0.27 and 0.73, respectively). Indeed, the EQE of the device with BDASBi is 2.7%, which is much higher than that of the device with PEBA (1.6%). They concluded that the OLED with BDASBi exhibits an increase in out-coupling efficiency by 45% due to the strongly horizontally oriented dipoles.

References

[1] H. Fukagawa, K. Watanabe, S. Tokito, *Organic Electronics*, **10**, 798–802 (2009).

[2] A. Kawakami, E. Otsuki, M. Fujieda, H. Kita, H. Taka, H. Sato, H. Usui, *Jpn. J. Appl. Phys.*, **47(2)**, 1279–1283 (2008); E. Otsuki, H. Sato, A. Kawakami, H. Taka, H. Kita, H. Usui, *Thin Solid Films*, **518**, 703 (2009).

[3] D. J. Milliron, I. G. Hill, C. Shen, A. Kahn, J. Schwartz, *J. Appl. Phys.*, **87(1)**, 572–576 (2000).

[4] C. C. Wu, C. I. Wu, J. C. Sturm, A. Kahn, *Appl. Phys. Lett.*, **70(11)**, 1348–1350 (1997).

[5] C. W. Tang, S. A. VanSlyke, C. H. Chan, *J. Appl. Phys.*, **65(9)**, 3610–3616 (1989).

[6] L. S. Hung, L. Z. Zheng, M. G. Mason, *Appl. Phys. Lett.*, **78(5)**, 673–675 (2001).

[7] C. Ganzorig, K.-J. Kwak, K. Yagi, M. Fujihira, *Appl. Phys. Lett.*, **79(2)**, 272–274 (2001).

[8] I. H. Campbell, J. D. Kress, R. L. Martin, D. L. Smith, N. N. Barashkov, J. P. Ferraris, *Appl. PLys. Lett.*, **71(24)**, 3528–3530 (1997).

[9] S. F. Hsu, C.-C. Lee, A. T. Hu, C. H. Chen, *Current Applied Physics*, **4**, 663–666 (2004); S.-F. Hsu, C.-C. Lee, S.-W. Hwang, H.-H. Chen, C. H. Chen, A. T. Hu, *Thin Solid Films*, **478**, 271–274 (2005).

[10] C. W. Chen, P. Y. Hsieh, H. H. Chiang, C. L. Lin, H. M. Wu, C. C. Wu, *Appl. Phys. Lett.*, **83(25)**, 5127–5129 (2003).

[11] L.-W. Chong, T.-C. Wen, Y.-L. Lee, T.-F. Guo, *Organic Electronics*, **9**, 515–521 (2008).

[12] S. A. VanSlyke, C. H. Chen, C. W. Tang, *Appl. Phys. Lett.*, **69(15)**, 2160–2162 (1996).

[13] Y. Shirota, T. Kobata, N. Noma, *Chem. Lett.*, **1145** (1989).

[14] Y. Shirota, Y. Kuwabara, H. Inada, T. Wakimoto, H. Nakada, Y. Yonemoto, S. Kawami and K. Imai, *Appl. Phys. Lett.*, **65(7)**, 807–809 (1994).

[15] S. H. Rhee, K. B. Nam, C. S. Kim, S. Y. Ryu, *ECS Solid State Lett.*, **3(3)**, R7-R10 (2014).

[16] S. Tokito, N. Noda and Y. Taga, *J. Phys. D-Appl. Phys.*, **29**, 2750–2753 (1996).

[17] Z. B. Deng, X. M. Ding, S. T. Lee, W. A. Gambling, *Appl. Phys. Lett.*, **74(15)**, 2227–2229 (1999).

[18] W. Hu, M. Matsumura, K. Furukawa, K. Torimitsu, *J. Phys. Chem. B*, **108**, 13116–13118 (2004).

[19] I-M. Chan, F. C. Hong, *Thin Solid Films*, **450**, 304–311 (2004).

[20] H. C. Im, D. C. Choo, T. W. Kim, J. H. Kim, J. H. Seo, Y. K. Kim, *Thin Solid Films*, **515** , 5099–5102 (2007).

[21] J. Li, M. Yahiro, K. Ishida, H. Yamada, K. Matsushige, *Synthetic Metals*, **151**, 141–146 (2005).

[22] T. Matsushima, Y. Kinoshita, H. Murata, *Appl. Phys. Lett.*, **91(25)**, 253504 (2007); T. Matsushima, H. Murata, *J. Appl. Phys.*, **104**, 034507 (2008).

[23] X. Zhou, M. Pfeiffer, J. Blochwitz, A. Werner, A. Nollau, T. Fritz, K. Leo, *Appl. Phys. Lett.*, **78(4)**, 410–412 (2001).

[24] C. W. Tang and S. A. VanSlyke, *Appl. Phys. Lett.*, **51**, 913 (1987).

[25] E. Han, L. Do, Y. Niidome and M. Fujihira, *Chem. Lett.*, **969** (1994).

[26] S. A. VanSlyke, C. H. Chen and C. W. Tang, *Appl. Phys. Lett.*, **69**, 2160 (1996).

[27] T. Noda, Y. Shirota, *Journal of Luminescence*, **87–89**, 1168–1170 (2000).

[28] Y. Shirota, K. Okumoto, H. Inada, *Synthetic Metals*, **111–112**, 387–391 (2000).

[29] K. Okumoto, K. Wayaku, T. Noda, H. Kageyama, Y. Shirota, *Synthetic Metals*, **111–112**, 473–476 (2000).

[30] C. Hosokawa, H. Higashi, H. Nakamura, T. Kusumoto, *Appl. Phys. Let.*, **67(26)**, 3853–3855 (1995); C. Hosokawa, H. Yokailin, H. Higashi, T. Kusumoto, *J. Appl. Phys.*, **78(9)**, 5831–5833 (1995).

[31] J. Kido, K. Hongawa, K. Okuyama and K. Nagai, *Appl. Phys. Lett.*, **64(7)**, 815–817 (1994).

[32] Y. Hamada, H. Kanno, T. Tsujioka, H. Takahashi, T. Usuki, *Appl. Phys. Lett.*, **75(12)** 1682–1684 (1999).

[33] C. W. Tang, S. A. VanSlyke, C. H. Chen, *J. Appl. Phys.*, **65**, 3610 (1989).

[34] C. Hosokawa, M. Eida, M. Matsuura, K. Fukuoka, H. Nakamura, T. Kusumoto, *Synth. Met.*, **91**, 3–7 (1997).

[35] Y. Kawamura, H. Kuma, M. Funahashi, M. Kawamura, Y. Mizuki, H. Saito, R. Naraoka, K. Nishimura, Y. Jinde, T. Iwakuma, C. Hosokawa, *SID 11 Digest*, 56.4 (p. 829) (2011).

[36] M. Kawamura, Y. Kawamura, Y. Mizuki, M. Funahashi, H. Kuma, C. Hosokawa, *SID 10 Digest*, 39.4 (p. 560) (2010).

[37] M. A. Baldo, D. F. O'Brien, Y. You, A. Shoustikov, S. Sibley, M. E. Thompson, and S. R. Forrest, *Nature*, **395**, 151 (1998).

[38] M. A. Baldo, S. Lamansky, P. E. Burrows, M. E. Thompson, S. R. Forrest, *Appl. Phys. Lett.*, **75**, 4–6 (1999).

[39] C. Adachi, M. A. Baldo, S. R. Forrest, *Appl. Phys. Lett*, **77**, 904–906 (2000).

[40] M. Ikai, S. Tokito, Y. Sakamoto, T. Suzuki, Y. Taga, *Appl. Phys. Lett.*, **79**, 156–158 (2001).

[41] C. Adachi, M. A. Baldo, M. E. Thompson, S. R. Forrest, *J. Appl. Phys.*, **90(10)**, 5048–5051 (2001).

[42] C. Adachi, R. C. Kwong, P. Djurovich, V. Adamovich, M. A. Baldo, M. E. Thompson, S. R. Forrest, *Appl. Phys. Lett.*, **79(13)**, 2082–2084 (2001).

[43] C. Adachi, M. A. Baldo, S. R. Forrest, S. Lamansky, M. E. Thompson, R. C. Kwong, *Appl. Phys. Lett.*, **78(11)**, 1622–1624 (2001).

[44] A. B. Tamayo, B. D. Alleyne, P. I. Djurovich, S. Lamansky, I. Tsyba, N. N. Ho, R. Bau, M. E. Thompson, *J. Am. Chem. Soc.*, **125**, 7377–7387 (2003).

[45] T. Yoshihara, Y. Sugiyama, S. Tobita, *Proc. of 6th Japanese OLED Forum*, S7–2 (p. 37) (2008).

[46] H. Sasabe, J. Takamatsu, T. Motoyama, S. Watanabe, G. Wagenblast, N. Langer, O. Molt, E. Fuchs, C. Lennartz, J. Kido, *Adv. Mater.*, **22**, 5003–5007 (2010).

[47] S. Tokito, T. Iijima, Y. Suzuri, H. Kia, T. Tsuzuki, F. Sato, *Appl. Phys. Lett.*, **83(3)**, 569–571 (2003);; S. Tokito, T. Tsuzuki, F. Sato, T. Iijima, *Current Appl. Phys.*, **5**, 331–336 (2005).

[48] I. Tanaka, Y. Tabata, S. Tokito, *Chem. Phys. Lett.*, **400**, 86–89 (2004).

[49] H. Sasabe, K. Minamoto, Y.-J. Pu, M. Hirasawa, J. Kido, *Organic Electronics*, **13**, 2615–2619 (2012).

[50] H. Uoyama, K. Goushi, K. Shizu, H. Nomura, C. Adachi, *Nature*, **492**, 234 (2012).

[51] A. Endo, K. Sato, K. Yoshimura, T. Kai, A. Kawada, H. Miyazaki, and C. Adachi, *Appl. Phys. Lett.*, **98**, 083302 (2011).

[52] Q. Zhang, J. Li, K. Shizu, S. Huang, S. Hirata, H. Miyazaki, C. Adachi, *J. Am. Chem. Soc.*, **134**, 14706–14709 (2012); bH. Tanaka, K. Shizu, H. Miyazaki, C. Adachi, *Chem. Commun.*, **48**, 11392–11394 (2012).

[53] C. Adachi, *Jpn. J. Appl. Phys.*, **53**, 060101 (2014).

[54] C. W. Tang and S. A. VanSlyke, *Appl. Phys. Lett.*, **51**, 913 (1987).

[55] Y.-J. Li, H. Sasabe, J.-J. Su, D. Takana, T. Takeda, Y.-J. Pu, J. Kido, *Chem. Lett.*, **38(7)**, 712–713 (2009).

[56] H. Sasabe, J. Kido, *Chem. Mater.*, **23**, 621–630 (2011).

[57] H. Sasabe, J. Kido, *Eur. J. Org. Chem.*, 7653–7663 (2013).

[58] C. W. Tang and S. A. VanSlyke, *Appl. Phys. Lett.*, **51**, 913–915 (1987).

[59] L. S. Hung, C. W. Tang, M. G. Mason, *Appl. Phys. Lett.*, **70(2)**, 152–154 (1997).; L. S. Hung, C. W. Tang, M. G. Mason, P. Raychaudhuri, J. Madathil, *Appl. Phys. Lett.*, **78(4)**, 544–546 (2001);M. G. Mason, C. W. Tang, L.-S. Hung, P. Raychaudhuri, J. Madathil, L. Yan, Q. T. Le, Y. Gao, S.-T. Lee, L. S. Liao, L. F. Cheng, W. R. Salaneck, D. A. dos Santos, J. L. Bredas, *J. Appl. Phys.*, **89(5)**, 2756–2765 (2001).

[60] G. E. Jabbour, B. Kippelen, N. R. Armstrong, N. Peyghambarian, *Appl. Phys. Lett.*, **73(9)**, 1185–1187 (1998).

[61] S. E. Shaheen, G. E. Jabbour, M. M. Morrell, Y. Kawabe, B. Kippelen, N. Peyghambarian, M.-F. Nabor, R. Schalaf, E. A. Mash, N. R. Armstrong, *J. Appl. Phys.*, **84(4)**, 2324–2327 (1998).

[62] T. Wakimoto, Y. Fukuda, K. Nagayama, A. Yokoi, H. Nakada, M. Tsuchida, *IEEE Transitions on Electron Devices*, **44(8)**, 1245–1248 (1997).

[63] A. R. Brown, D. D. C. Bradley, J. H. Burroughes, R. H. Friend, N. C. Greenham, P. L. Burn, A. B. Holmes, A. Kraft, *Appl. Phys. Lett.*, **62(23)**, 2793–2795 (1992).

[64] I. D. Parker, *J. Appl. Phys.*, **75(3)**, 1656–1666 (1994).

[65] R. H. Friend, R. W. Gymer, A. B. Holmes, J. H. Burroughes, R. N. Marks, C. Taliani, D. D. C. Bradley, D. A. Dos Santos, J. L. Bredas, M. Lögdlung, W. R. Salaneck, *Nature*, **397**, 121–128 (1999).

[66] Y. Cao, G. Yu, I. D. Parker, A. Heeger, *J. Appl. Phys.*, **88(6)**, 3618–3623 (2000).

[67] T. M. Brown, R. H. Friend, I. S. Millard, D. L. Lacey, T. Butler, J. H. Burroughes, F. Cacialli, *J. Appl. Phys.*, **93(10)**, 6159–6172 (2003).

[68] C. W. Chen, P. Y. Hsieh, H. H. Chiang, C. L. Lin, H. M. Wu, C. C. Wu, *Appl. Phys. Lett.*, **83(25)**, 5127–5129 (2003).

[69] K. Okamoto, Y. Fujita, Y. Ohnishi, S. Kawato, M. Koden, *Proc. of 7th Japanese OLED Forum*, S9-2 (2008).

[70] J. Kido, K. Nagai, Y. Okamoto, *IEEE Transactions on Electron Devices*, **40(7)**, 1342–1344 (1993).

[71] J. Kido, T. Matsumoto, *Appl. Phys. Lett.*, **73**, 2866–2868 (1998).

[72] K. Walzer, B. Maennig, M. Pfeiffer, K. Leo, *Chem. Rev.*, **107**, 1233–1271 (2007).

[73] J. Endo, J. Kido, T. Matsumoto, *Ext. Abst. (59th Autumn Meet. 1998); Jpn. Soc. Appl. Phys.*, 16a-YH-10 (p. 1086) (1998).

[74] J. Endo, T. Matsumoto, J. Kido, *Jpn. J. Appl. Phys*, **41**, L800–L803 (2002).

[75] C. Schmitz, H.-W. Schmidt, M. Thelakkat, *Chem. Mater.*, **12**, 3012–3019 (2000).

[76] Y.-J. Pu, M. Miyamoto, K. Nakayama, T. Oyama, M. Yokoyama, J. Kido, *Org. Electron*, **10**, 228 (2009).

[77] C. Adachi, R. C. Kwong, P. Djurovich, V. Adamovich, M. A. Baldo, M. E. Thompson, S. R. Forrest, *Appl. Phys. Lett.*, **79(13)**, 2082–2084 (2001).

[78] R. C. Kwong, M. R. Nugent, L. M. Michalski, T. Ngo, K. Rajan, Y.-J. Tung, M. S. Weaver, T. X. Xhou, M. Hack, M. E. Thompson, S. R. Forrest, J. J. Brown, *Appl. Phys. Lett.*, **81(1)**, 162–164 (2002).

[79] V. I. Adamovich, S. R. Cordero, P. I. Djurovich, A. Tamayo, M. E. Thompson, B. W. D'Andrade, S. R. Forrest, *Organic Electronics*, **4**, 77–87 (2003).

[80] M. B. Khalifa, D. Vaufrey, J. Tardy, *Organic Electronics*, **5**, 187–198 (2004).

[81] M. A. Baldo, S. Lamansky, P. E. Burrows, M. E. Thompson, S. R. Forrest, SR, *Appl. Phys. Lett.*, **75(1)**, 4–6 (1999).

[82] C. Adachi, M. A. Baldo, S. R. Forrest, S. Lamansky, M. E. Thompson, R. C. Kwong, *Appl. Phys. Lett.*, **78(11)**, 1622–1624 (2001).

[83] M. Fujihira, C. Ganzorig, *Materials Science and Engineering*, **B85**, 203–208 (2001).

[84] D. Gebeyehu, K. Walzer, G. He, M. Pfeiffer, K. Leo, J. Brandt, A. Gerhard, P. Stößel, H. Vestweber, *Synthetic Metals*, **148**, 205–211 (2005).

[85] H. Ikeda, J. Sakata, M. Hayakawa, T. Aoyama, T. Kawakami, K. Kamata, Y. Iwaki, S. Seo, Y. Noda, R. Nomura, S. Yamazaki, *SID 06 Digest*, P-185 (p. 923) (2006).

[86] S.-H. Su, C.-C. Hou, J.-S. Tsai, M. Yokoyama, *Thin Solid Films*, **517**, 5293–5297 (2009).

[87] J. H. Burroughes, D. D. Bradley, A. R. Brown, R. N. Markes, K. Mackay, R. H. Friend, P. L. Burns and A. B. Holmes, *Nature*, **347**, 539–541 (1990).

[88] R. H. Friend, R. W. Gymer, A. B. Holmes, J. H. Burroughes, R. N. Marks, C. Taliani, D. D. C. Bradley, D. A. Dos Santos, J. L. Bredas, M. Logdlund, W. R. Salaneck, *Nature*, **397**, 121–128 (1999).

[89] D. Braun, A. Heeger, *Appl. Phys. Lett.*, **58**, 1982 (1991).

[90] P. L. Burn, A. B. Holmes, A. Kraft, D. D. C. Bradley, A. R. Brown, R. H. Friend, R. W. Gymer, *Nature*, **356**, 47–49 (1992).

[91] M. Fukuda, K. Sawada, S. Morita, K. Yoshino, *Synth. Met.*, **41**, 855 (1991).

[92] Y. Ohmori, M. Uchida, K. Muro, K. Yoshino, *Jpn. J. Appl. Phys.*, **30**, L1941 (1991).

[93] M. T. Bernius, M. Inbasekaran, J. O'Brien, W. Wu, *Adv. Mater.*, **12(23)**, 1737–1750 (2000).

[94] Y. Yang, E. Westerweele, C. Zhang, P. Smith, A. J. Heeger, *J. Appl. Phys.*, **77(2)**, 694–698 (1995).

[95] Y. Cao, G. Yu, C. Zhang, R. Menon, A. J. Heeger, *Synth. Met.*, **87(2)**, 171–174 (1997).

[96] S. A. VanSlyke, C. H. Chen, C. W. Tang, *Appl. Phys. Lett.*, **69**, 2160 (1996).

[97] G. Greczynski, T. Kugler, W. R. Salaneck, *Thin Solid Films*, **354**, 129–135 (1999).

[98] A. Elschner, F. Bruder, H. W. Heuer, F. Jonas, A. Karbach, S. Kirchmeyer, S. Thurm, *Synth. Met.*, **111**, 139–143 (2000).

[99] A. Elschner, F. Jonas, S. Kirchmeyer, K. Wussow, *Proc. AD/IDW'01*, OEL3-3 (2001).

[100] R. H. Friend, *Proc. AM-LEC'01*, OLED-1 (2001).

[101] X. Gong, D. Moses, A. J. Heeger, S.Liu, A. K. –Y. Jen, *Appl. Phys. Lett.*, **83(1)**, 183–185 (2003).

[102] M. Leadbeater, N. Patel, B. Tierney, S. O'Connor, I. Grizzi, C. Town, *SID 04 Digest*, 11.5L (p. 162) (2004).

[103] J. Morgado, R. H. Friend, F. Cacialli, *Appl. Phys. Lett.*, **80(14)**, 2436–2438 (2002).

[104] N. Conway, C. Foden, C. Roberts, I. Grizzi, *Proc. Euro Display*, 18.4 (p. 492) (2005).

[105] T. Shirasaki, T. Ozaki, K. Sato, M. Kumagai, M. Takei, T. Toyama, S. Shinoda, T. Tano, R. Hattori, *SID 04 Digest*, 57.4L (p. 1516) (2004).

[106] Y. Fujita, M. Koden, SID/MAC OLED Research and Technology Conference (Park Ridge, USA, 2004); Y. Fujita, A. Tagawa, M. Koden, *OLEDs Asia 2005* (2005).

[107] Y. Hatanaka, Y. Fujita, M. Koden, *Proc.of 2nd Japanese OLED Forum*, S7–2 (p. 53) (2006).

[108] S. Sakamoto, M. Okumura, Z. Zhao, Y. Furukawa, *Chem. Phys. Lett.*, **412**, 395–398 (2005).

[109] T.-F. Guo, S.-C. Chang, Y. Yang, R. C. Kwong, M. E. Thompson, *Organic Electronics*, **1**, 15–20 (2000).

[110] C. L. Lee, K. B. Lee, J. J. Kim, *Appl. Phys. Lett.*, **77(15)**, 2280–2282 (2000); C. L. Lee, K. B. Lee, J. J. Kim, *Mater. Sci. Eng.*, **B85**, 228–231 (2001).

[111] S. Lamansky, R. C. Kwong, M. Nugent, P. I. Djurovich, M. E. Thompson, *Organic Electronics*, **2(1)**, 53–62 (2001).

[112] P. A. Lane, L. C. Palilis, D. F. O'Brien, C. Giebeler, A. J. Cadby, D. G. Lidzey, A. J. Campbell, W. Blau, D. D. C. Bradley, *Phys. Rev. B.*, **63(23)**, 235206 (2001): D. F. O'Brien, C. Giebeler, R. B. Fletcher, A. J. Cadby, L. C. Palilis, D. G. Lidzey, P. A. Lane, D. D. C. Bradley, W. Blau, *Synthetic Metals*, **116**, 379–383 (2001).

[113] K. M. Vaeth, and C. W. Tang, *J. Appl. Phys.*, **92(7)**, 3447–3453 (2002).

[114] Y. Hino, H. Kajii, Y. Ohmori, *Organic Electronics*, **5**, 265–270 (2004).

[115] C. L. Lee, N. G. Kang, Y. S. Cho, J. S. Lee, J. J. Kim, *Optical Materials*, **21**, 119–123 (2002).

[116] S. Tokito, M. Suzuki, F. Sato, M. Kamachi, K. Shirane, *Organic Electronics*, **4**, 105–111 (2003).

[117] M. Tokito, M. Suzuki, F. Sato, *Thin Solid Films*, **445**, 353–357 (2003).

[118] M. Suzuki, M. Tokito, F. Sato, T. Igarashi, K. Kondo, T. Koyama, T. Yamaguchi, *Appl. Phys. Lett.*, **86(10)**, 103507 (2005).

[119] M. Halim, J. N. G. Pillow, I. D. W. Samuel, P. L. Burn, *Adv. Mater.*, **11(5)**, 371–374 (1999); M. Halim, I. D. W. Samuel, J. N. G. Pillow, P. L. Burn, *Synthetic Metals*, **102**, 1113–1114 (1999).

[120] J. M. Lupton, I. D. W. Samuel, R. Beavington, P. L. Burn, H. Bässler, *Adv. Mater.*, **13(4)**, 258–261 (2001).

[121] J. M. Lupton, I. D. W. Samuel, M. J. Frampton, R. Beavington, P. L. Burn, *Adv. Funct. Mater.*, **11(4)**, 287–294 (2001).

[122] J. P. J. Markham, T. Anthopoulos, S. W. Magennis, I. D. W. Samuel, N. H. Male, O. Salata, S.-C. Lo, P. L. Burn, *SID 02 Digest*, L-8 (p. 1032) (2002).

[123] J. Pillow, Z. Liu, C. Sekine, S. Mikami, M. Mayumi, *SID 2005 Digest*, 22.4 (p. 1071) (2005); C. Sekine, S. Mikami, M. Mayumi, Y. Akino, H. Onishi, J. Pillow, Z. Liu, *Proc of 1st Japanese OLED Forum*, S3–1 (2005).

[124] T. Tsuzuki, N. Shirasawa, T. Suzuki, S. Tokito, *Jpn. J. Appl. Phys.*, **44(6A)**, 4151–4154 (2005).

[125] N. Iguchi, Y.-J. Pu, K. Nakayama, M. Yokoyama, J. Kido, *Organic Electronics*, **10**, 465–472 (2009).

[126] J. W. Levell, J. P. Gunning, P. L. Burn, J. Robertson, I. D. W. Samuel, *Organic Electronics*, **11**, 1561–1568 (2010).

[127] N. Herron, W. Gao, *SID 10 Digest*, 32.3 (p. 469) (2010).

[128] D. Yokoyama, A. Sakaguchi, M. Suzuki, C. Adachi, *Organic Electronics*, **10**, 127–137 (2010).

[129] J. Frischeisen, D. Yokoyama, A. Endo, C. Adachi, W. Brütting, *Organic Electronics*, **12**, 809–817 (2011).

5

OLED Devices

Summary

While the fundamental device structure of OLEDs is simple in a sense, there are various types of OLED devices, which can be classified into various categories. From the point of view of emitting directions, OLED devices are classified as bottom emission, top emission, and both-side emission (transparent). OLED devices are also classified into normal and inverted structures from the point of view of the stacking order of the electrodes.

In addition, this chapter describes some useful technologies for practical OLED displays and lighting: white OLEDs, full-color technologies, micro-cavity structures, multi-photon structures, and encapsulating technologies.

Key words

device, bottom emission, top emission, both-side emission, transparent, white, full-color, micro-cavity, multi-photon, encapsulation

5.1 Bottom Emission, Top Emission, and Transparent Types

OLED devices are classified as bottom emission, top emission, and both-side emission (transparent) from the emitting directions. These are illustrated in Fig. 5.1, and Table 5.1 shows a comparison between bottom emission and top emission OLED devices.

The bottom emitting OLED device is, in a sense, a general and classical structure. In most cases, an anode is deposited on a substrate and then organic layers and a cathode are deposited in that order. The emission can be observed through the substrate. Therefore, substrates and anodes in bottom emitting OLED devices should be transparent. On the other hand, opaque and reflective cathodes can be used. In addition, opaque desiccant can be set for encapsulation

OLED Displays and Lighting, First Edition. Mitsuhiro Koden.

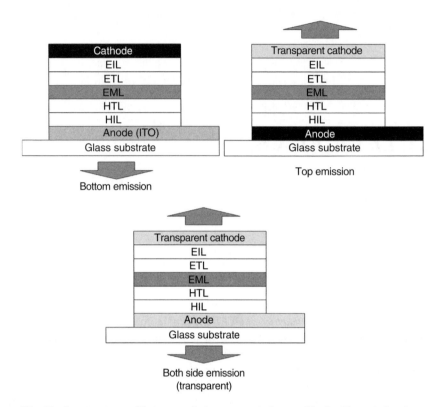

Figure 5.1   Device structures of bottom emission, top emission, and both-side emission (transparent) OLEDs

Table 5.1   Comparison between bottom and top emission OLED devices

	Bottom emission	Top emission
Substrate	transparent	no restriction
Upper electrode (normally cathode)	opaque	transparent
Encapsulation	rather easy	not so easy (transparency required)
Production	rather easy	not so easy
Aperture ratio in AM-OLED	low	high
Lifetime	short	long
High resolution	difficult	possible

as shown in Fig. 5.2. For these reasons, it is fairly easy to fabricate bottom emitting OLED devices, comparing to top emitting OLED devices. We can utilize these easy fabrication processes where ITO, organic layers, opaque cathodes such as Al and Ag, are sequentially deposited on a glass substrate, followed by an encapsulating process with a desiccant.

However, bottom emitting OLED devices have several restrictions in the application to active-matrix OLED displays. As described in Chapter 8, active-matrix OLED displays require multiple TFTs (thin film transistors) for each pixel. Therefore, the aperture ratio of TFT substrates tends to be small. Even with a resolution of about 200 ppi, the aperture ratios tend to be smaller than 30%, if the OLED device structure is bottom emitting. Considering the fact that current mobile displays need extremely high resolution, such as 400–500 ppi or more, the restriction of aperture ratio is considered to be a severe limitation. Since this low aperture ratio requires OLED devices to have high luminance in order of achieve a designed luminance, the elevated luminance gives rise to a drastic reduction in lifetime.

The top emitting OLED device structure can intrinsically solve these serious issues of the bottom emitting OLED device. Since the emission is observed through the cathode in top emitting OLED devices, as shown in Fig. 5.1, the aperture ratio is essentially independent of the TFT circuits in active-matrix OLED displays. From the point of view of TFT circuit design, we can utilize the whole area of each pixel for designing TFT circuits and bus lines (row and column electrode lines), obtaining larger aperture ratio than the corresponding bottom emitting OLED device structure. The increase in aperture ratio by adopting top emitting device structure can contribute to elongation of lifetime. By adopting a top emitting device structure, high resolution (such as 400 ppi or more) mobile AM-OLED displays can be fabricated.

Top emitting OLED devices have been studied and reported in the early R&D stage. In 1994, Baigent et al. of the Cambridge group had already reported top emission OLEDs fabricated on a silicon substrate [1]. They deposited a thin layer of aluminum, a layer of poly(cyanoterephthalylidene), CN-PPV, and a layer of poly(p-phenylene vinylene), PPV. On the polymer layers, ITO was deposited by sputtering. This structure is not only a top emitting device but also an inverted structure, which is described in the next section.

As shown in Fig. 5.1, in top emitting devices, bottom electrodes (normally anodes) should be opaque and reflective, and top electrodes (normally cathodes) should be transparent. In addition, opaque desiccant cannot be used. Therefore, transparent encapsulating technologies as shown in Fig. 5.2 are necessarily required. For these reasons, it is more difficult to fabricate top emitting OLED devices than bottom emitting OLED devices.

Top emission OLED devices require a reflective bottom electrode (normally anode), while bottom emission OLEDs usually use a transparent ITO anode. Typical reflective anodes for top emission OLED devices are Ag, Ag/Ag$_2$O [2], and Ag/ITO [3].

The transparent upper electrode (normally cathode) is one of the issues in top emission OLEDs. While one of typical transparent cathodes is ITO, the sputtering of ITO tends to cause damage to OLED devices. To overcome this issue, novel ITO deposition technologies with little or no damage have been developed, for example mirrortron sputtering and facing targets sputtering. Another candidate for transparent cathodes is LiF/Al/Ag [4].

Since top emitting OLED devices usually use semi-transparent cathodes, top emitting OLED devices tend to show a micro-cavity effect. By combining the micro-cavity structure in top emitting devices, it is possible to improve color purity and efficiency. Therefore, in top emitting OLED devices, OLED architecture based on micro-cavity effect is often required. A typical structure of top emitting OLED devices with the micro-cavity technology is shown in Fig. 5.3 [5]. The micro-cavity technology is described in Section 5.5.

The third type of OLED device is the both-side emitting OLED. In this case, emission is observed from both sides, and the device is transparent, as shown in Fig. 5.1. Therefore, in OLED fields, both-side emitting OLED device is usually called transparent OLED (TOLED) [6]. TOLEDs require transparent anode and transparent cathode. Since TOLED is a modified technology of top emitting OLED, TOLEDs have the same issues as top emitting OLEDs.

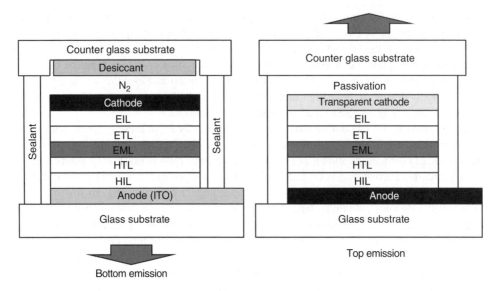

Figure 5.2   Typical encapsulations in bottom emitting and top emitting OLEDs

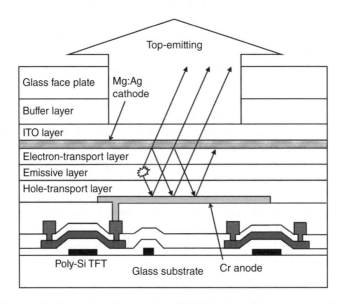

Figure 5.3   A typical example of top emitting OLED device with the micro-cavity effect [5]

Using top emission OLED technologies, transparent OLEDs can be fabricated. In 1996, the group of Forrest and Thompson reported a transparent OLED using an ITO anode and a top electrode composed of a very thin layer of Mg:Ag and an overlaying ITO layer [6]. The device structure is shown in Fig. 5.4.

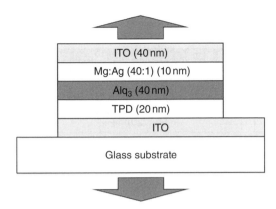

Figure 5.4 Schematic diagram of the TOLED structure reported by Bulovic et al. [6]

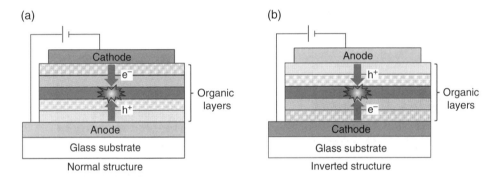

Figure 5.5 Normal and inverted structures of OLEDs

5.2 Normal and Inverted Structures

In usual OLEDs, the bottom electrode on the substrate is the anode. This structure is referred to as a normal structure. On the other hand, inverted structure is possible, where the bottom electrode is the cathode. These two structures are shown in Fig. 5.5. While normal structure is usually used, there are only few reports on inverted structure. Some examples of the inverted structure are described below [1, 7–12].

As described in Section 5.1, the top emitting OLED device reported by Baigent et al. had an inverted OLED device structure [1]. In this case, a cathode is deposited on a substrate and an anode is finally deposited on organic layers.

Morii et al. of Seiko Epson, also reported a top emitting inverted OLED device, in which a light emitting polymer (F8BT) is sandwiched between TiO_2 and MoO_3 [7]. The device structure is shown in Fig. 5.6. The TiO_2 acts as electron injector and the MoO_3 provides hole injection. The OLED device showed lower turn-on voltage and higher air stability than conventional polymer OLED devices.

Meng et al. of Jilin University (PR China) reported an inverted top emitting OLED, using MoO_x/Ag anode as a top electrode [8].

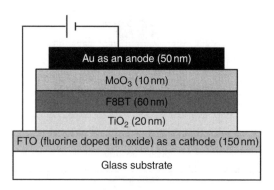

Figure 5.6 Device structure of an OLED with top emitting inverted structure reported by Morii et al. [7]

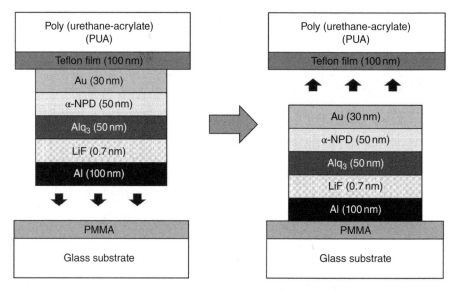

Figure 5.7 Schematic illustration of the inverted top emitting ITOLED by whole device transfer method reported by Kim et al. [9]

Kim et al. of Seoul National University (Korea) fabricated top emitting OLEDs with an inverted structure using a transfer method [9]. They fabricated a normal OLED structure on a poly(urethane-acrylate) (PUA) substrate with a thin Teflon film (AF2400, DuPont, ~100 nm). After fabricating a bottom emitting OLED device, the whole device is brought into contact with another substrate, as shown in Fig. 5.7. They also reported on fabricating a flexible inverted top emitting OLED device by using this transfer method.

Noh et al. of LG Chem Research Park (Korea) fabricated red and green phosphorescent OLED devices with the top emitting inverted structure, reporting that the inverted structure provided higher current efficiency and lower driving voltage than those of the corresponding normal structure [10]. They also reported that a current efficiency of 93.3 cd/A (at 4.6 V and 1000 cd/m²) was obtained in the green phosphorescent OLED device with the inverted structure in spite of the absence of any light extraction enhancement technology.

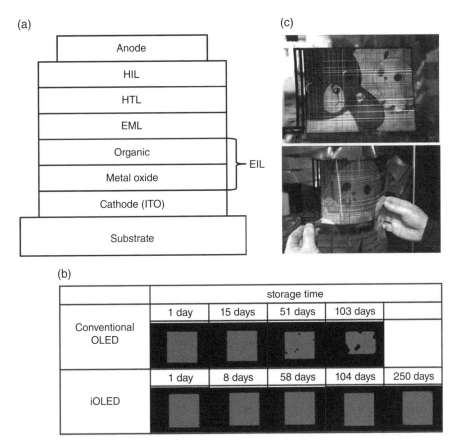

Figure 5.8 Inverted OLED device reported by Fukagawa et al. [12]. (a) Device structure. (b) Images of light-emitting areas of OLEDs as a function of storage time. The emitting area of the non-degraded OLED is 3×3 mm². (c) Photographs of fabricated flexible AM-OLED display with the inverted OLED structure driven by IGZO-TFTs

On the other hand, Lee et al. of Korea University, etc. (Republic of Korea) reported a bottom emitting inverted OLED devices with glass/ITO/Al/Liq/Alq₃/α-NPD/WO₃/Al, investigating several cathode materials [11]. In this case, the bottom cathode electrode is very thin (1.5 nm). They reported that Al/Liq works as an efficient electron injection layer.

For a display application, Fukagawa et al. of Japan Broadcasting Corporation (NHK) fabricated an 8″ VGA flexible AM-OLED display using an inverted bottom emitting OLED device structure [12], also reporting that the inverted OLED shows better storage lifetime than the conventional OLED, due to the better stability against attack by oxygen and moisture as shown in Fig. 5.8.

5.3 White OLEDs

White OLED emission was scientifically first reported by Kido et al. of Yamagata University (Japan) in 1994 [13]. They doped three fluorescent dyes with different emission spectra to the emission layer consisting of poly(n-vinylcarbazole) (PVK), obtaining white emission.

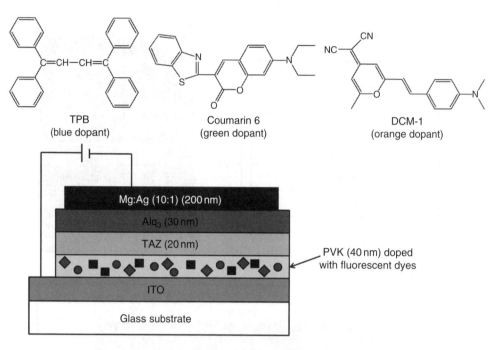

TPB
(blue dopant)

Coumarin 6
(green dopant)

DCM-1
(orange dopant)

Mg:Ag (10:1) (200 nm)

Alq$_3$ (30 nm)

TAZ (20 nm)

PVK (40 nm) doped
with fluorescent dyes

ITO

Glass substrate

Figure 5.9   Schematic device structure of a white OLED using a PVK film doped with fluorescent dyes with different colors [13]

The schematic device structure is shown in Fig. 5.9. The hole-transporting emission layer consists of PVK doped with three fluorescent dyes, which are blue-emitting 1,1,4,4-tetraphenyl-1,3-butadiene (TPD), green-emitting coumarin 6, and orange-emitting DCM-1. On the PVK layer, they deposited two organic layers of 1,2,4-triazole derivative (TAZ) and tris(8-quinolinolato)aluminum(III) complex (Alq) and Mg:Ag cathode. In this device, they observed white emission with 3400 cd/m² at a drive voltage of 14 V, while the power efficiency was still 0.83 lm/W.

While the first attempt at white OLED consisted of the host polymer layer doped with fluorescent dyes of different colors, practical white OLED architecture is the stacked emission layers with different colors, as shown in Fig. 5.10. In the early developmental stage, the stacking of blue and orange emission layers was often used. At present, the concept of the multi-layer structure for white emission has led current practical white emission OLED architectures.

The first attempt at such stacked emission layers was reported by Kido et al. of Yamagata University (Japan) in 1995 [14]. Their idea for obtaining white emission was to control the carrier recombination zone so that the recombination takes place in different layers. Based on this idea, Kido et al. stacked three RGB emitter layers with different carrier transport properties. The device structure is glass/ITO/TPD(40 nm)/p-EtTAZ(3 nm)/Alq$_3$(5 nm)/Alq$_3$(5 nm) (doped with 1 mole % of Nile red)/Alq$_3$(40 nm)/Mg:Ag(10:1). A hole-transporting triphenyldiamine derivative (TPD) gives blue emission with emission peaks at around 410–420 nm. A hole-blocking 1,2,4-triazole derivative (p-EtTAZ) provides electron transport and hole blocking. An electron-transporting aluminum complex (Alq$_3$) gives green emission with an emission peak of 520 nm. In the next layer, Nile red with an emission peak of 600 nm is doped

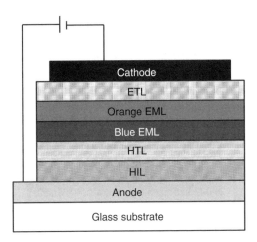

Figure 5.10 White OLED architecture with stacking multiple emission layers with different colors

with Alq$_3$, giving red emission. The device showed three emission peaks of 410–420, 520, and 600 nm, corresponding to emission from TPD, Alq$_3$, and Nile red. They reported white emission with luminance better than 2000 cd/m^2 at a drive voltage of 15–16 V.

After the report by Kido et al., various materials and structures for multi-layer white OLEDs have been studied and developed. Since the efficiency of OLEDs with combinations of RGB fluorescent materials is limited, several approaches for obtaining highly efficient white OLEDs have been investigated, mainly utilizing phosphorescent materials.

One useful concept is to combine a blue fluorescent material and phosphorescent materials for the other colors, due to the fact that green and red phosphorescent materials have the potential for practical uses but blue ones do not. This approach is a so-called hybrid white OLED.

Sun et al. of Forrest's group at Princeton University (USA) reported such hybrid white OLEDs with the maximum EQE of 18.7±0.5 % and a power efficiency of 37.6±0.6 lm/W, using a blue fluorescent material in combination with green and red phosphorescent materials [15]. The device structure is shown in Fig. 5.11. As the design concept, a blue fluorescent emitter harnesses all the electrically generated high energy singlet excitons for blue emission and phosphorescent emitters harvest the remainder of the low-energy triplet excitons for green and red emission.

The other approach is a phosphorescent multiple emissive layer.

Tokito et al. in NHK Science and Technical Research Laboratories and Tokyo University of Science (Japan) developed white phosphorescent OLEDs with greenish-blue and red emitting layers [16, 17]. The device structure and the molecular structure utilized in the device are shown in Fig. 5.12. (CF$_3$ppy)$_2$Ir(pic) is a greenish-blue phosphorescent material and (btp)$_2$Ir(acac) is a red phosphorescent material. They inserted a 3 nm BAlq layer as an exciton-blocking layer between the two phosphorescent emitting layers. They also deposited a 45 nm BAlq layer as an electron-transporting and hole-blocking layer between the red emitting layer and the cathode. They achieved white emission with a maximum quantum efficiency of 12%, a luminance efficiency of 18 cd/A, and a maximum power efficiency of 10 lm/W at the current density of 0.01 mA/cm^2.

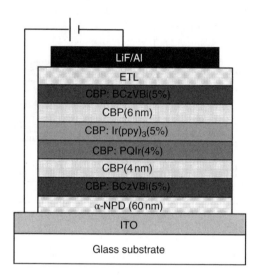

Figure 5.11 Device structure of a white OLED device with fluorescent and phosphorescent emissive layers reported by Sun et al. [15]

Cheng et al. of Jilin University, etc. (China) reported a phosphorescent white OLED in which a blue emissive layer is sandwiched between red and green layers [18]. Their device structure is shown in Fig. 5.13. While the achieved maximum power efficiency is not so high (9.9 lm/W), they achieved a high color rendering index of 82.

Reineke et al., of Leo's group at the Institute für Angewandte Photophysik (Germany), reported very high fluorescent tube efficiency of 81 lm/W in white OLEDs with a phosphorescent red-green-blue (RGB) multiple emissive layer and light-out-coupling enhancement techniques, such as high refractive index substrates and half-spheres [19].

For stacking emission layers with different colors, an useful alternative approach is to use the multi-photon structure (tandem structure), which is described in Section 5.6.

Phosphorescent triplet-doped emissive layer is another approach. This technology uses the fact that excited energies are easily transferred from high energy blue dopants to green ones and then red ones. Therefore, in order to obtain three color emissions with a reasonable balance, the doping concentrations of dopants have to be carefully optimized in the order of $B > G \gg R$. In general, the concentration of the red dopant is lower than 1%.

D'Andrade et al. of Forrest's group at Princeton University (USA) reported a white OLED device in which three color phosphorescent dopants were doped to a host material in the emission layer [20]. The energy level diagram of their white phosphorescent OLED device is shown in Fig. 5.14. In this device, three color phosphors, PQIr, Ir(ppy)3, and Fir6 were co-doped into a wide energy gap UGH4 host. This device gave 42 lm/W.

5.4   Full-Color Technology

The realization of full-color images is strongly demanded for electronic displays. At present, most electronic displays in our lives are full-color displays as is obvious in TVs, mobile phones, smart phones, personal computers, desktop computers, digital cameras,

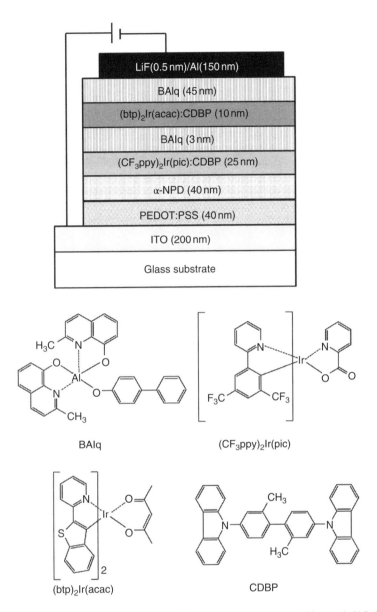

Figure 5.12 Schematic structure of a white OLED device developed by Tokito et al. [16, 17], accompanied by the molecular structures of the materials used

signboards, information displays, etc. Therefore, it can be said that full-color technologies are essentially required.

In electronic displays, full color is usually realized by the arrangement of red, green, and blue (RGB) sub-pixels, using a principle of additive combination of primary colors. Figure 5.15 shows typical technologies for realizing such an arrangement in OLED devices. They are RGB-side-by-side, white emission with color filter and blue emission with color changing medium.

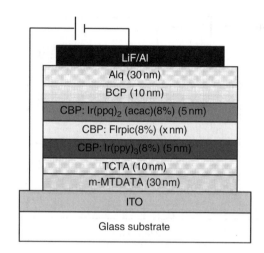

Figure 5.13 Structure of a white OLED device with a phosphorescent multiple emissive layer reported by Cheng et al. [18]

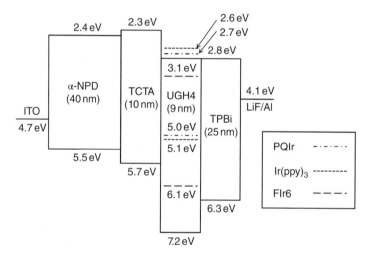

Figure 5.14 The energy level diagram of a white phosphorescent OLED device in which three color phosphorescent dopants were doped to a host material in the emission layer reported by D'Andrade et al. [20]

In a sense, the most fundamental method is the RGB-side-by-side method, in which emitting sub-pixels of different colors are fabricated. The advantage of this method is to utilize the performance of OLED device itself. However, this method requires additional technologies for fabricating RGB-side-by-side structures. In vacuum evaporation OLEDs, the mask deposition technique is often applied. This is the most common technology but it is difficult to be apply to large size displays and high resolution displays due to the difficulty of mask treatment in production. In solution process OLEDs, printing technologies such as ink-jet and relief printing can be used for fabricating RGB-side-by-side structures. Laser patterning technologies have also been presented. While these process technologies will be described in Chapter 6, they still have several issues.

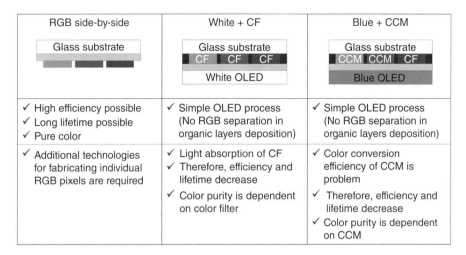

RGB side-by-side	White + CF	Blue + CCM
Glass substrate	Glass substrate CF | CF | CF White OLED	Glass substrate CCM | CCM | CF Blue OLED
✓ High efficiency possible ✓ Long lifetime possible ✓ Pure color	✓ Simple OLED process (No RGB separation in organic layers deposition)	✓ Simple OLED process (No RGB separation in organic layers deposition)
✓ Additional technologies for fabricating individual RGB pixels are required	✓ Light absorption of CF ✓ Therefore, efficiency and lifetime decrease ✓ Color purity is dependent on color filter	✓ Color conversion efficiency of CCM is problem ✓ Therefore, efficiency and lifetime decrease ✓ Color purity is dependent on CCM

Figure 5.15 Technologies for obtaining RGB colors for full-color OLED devices

The combination of white emission OLED with color filter is also often used. Since this method requires no RGB separating pixels, the fabrication is fairly easy. However, since color filters absorb light from white emission, efficiency and lifetime are restricted. As a counter-technology, four-pixel structures using RGBW (RGB + white) have been proposed [21].

Another approach is a combination of a blue emission with a color-changing medium (CCM) [22–25]. Since this method requires no RGB separating pixels, the fabrication process is also fairly easy. However, since the conversion efficiency is not 100%, OLED efficiency and lifetime are restricted. At present, this technology is scarcely used.

While there are other methods such as stacked RGB cell [26] and color changed emitters by photo-breaching [27] for obtaining full color, they are also rarely used at present.

5.4.1 RGB-Side-by-Side

For fabricating OLED devices with RGB-side-by-side, various technologies have been proposed and investigated.

In vacuum evaporation methods, fine metal mask (FMM) deposition is the most generally used method. In solution processes, printing techniques such as ink-jet are often used. In addition, laser patterning methods of evaporated or coated film have also been proposed and investigated.

These are described in Chapter 6 in detail.

5.4.2 White + CF

White emission and color filter is a useful technology for full-color OLED displays, being applied to various OLED displays.

Mameno et al. in Sanyo Electric (Japan) developed an AM-OLED display using white emitter and color filter [28]. The display has a bottom emitting OLED device structure and the specifications are 2.5″ diagonal, 240 × 320 dots.

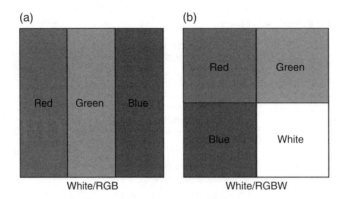

Figure 5.16   Examples of pixels in W-RGB and W-RGBW

Kashiwabara et al. at Sony developed an AM-OLED display with a top emitting device structure, using white emitter and color filter array [29]. The display utilizes the micro-cavity effect for obtaining high color purity. The specifications are 12.5″ diagonal, 854×480 dots.

As a modified technology of white emitter and color filter (W-RGB), four-pixel structures using RGB and white pixels (W-RGBW) have been proposed and investigated [21]. Examples of pixel layouts of color filters in W-RGB and W-RGBW are shown in Fig. 5.16. In the W-RGB method, a large part of white emitted light is absorbed by each RGB color filter. Therefore, it is difficult to obtain high efficiency. On the other hand, in the white pixel in the W-RGBW method, no white emitted light is absorbed, making possible to achieve high efficiency.

Since most of colors in actual display images include a component of white, the W-RGBW can realize much lower power consumption than the W-RGB method. Indeed, they evaluated the power consumption of the various actual images in the W-RGB and W-RGBW methods, concluding that the power consumption in the W-RGBW is almost half of that in the W-RGB [21].

By using the W-RGBW method, Tsujimura et al. successfully fabricated an 8.1″ prototype AM-OLED display (WVGA, 300 cd/m², LTPS-TFT) with 100% NTSC color gamut, also developing RGBW rendering algorithm [30].

5.4.3   Blue Emission with Color Changing Medium (CCM)

As described above, the combination of a blue emission with color changing mediums (CCM) has the advantage of no requirement for RGB patterning in OLED displays [22–25].

The device mechanism is shown in Fig. 5.17. The OLED device is monochrome blue emitted OLED. In the red and green pixels, the emitted blue light is absorbed in the CCM layer. Then, the CCM layer emits each color, according to the contained phosphor. On the other hand, in the blue pixel, the color filter is often combined in order to obtain deep blue color.

Hosokawa et al. of Idemitsu Kosan Co., Ltd (Japan) reported the detail of this technology [22–25]. The CCM layer is made by organic fluorescent mediums which change color from blue to green or red. In the blue pixel, a blue color filter is fabricated instead of CCM in order to improve the blue color purity. The CCM is patterned by a photolithographic process.

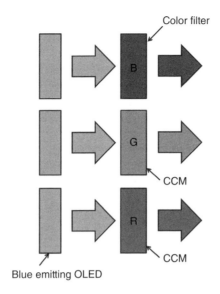

Figure 5.17   The device mechanism of the CCM

For the green pixels, a photoresist polymer containing green fluorescent dyes is coated and then patterned by a photolithographic process. For the red pixels, a transparent resin containing red fluorescent dyes is coated and then patterned by an etching process through a photoresist mask. They fabricated a CCM structure with sub-pixel patterns of 300 μm pitch and fabricated a full-color PM-OLED display.

5.5   Micro-Cavity Structure

Micro-cavity structure is often used in OLED devices, particularly in top emitting OLED devices. In OLED devices, the micro-cavity effect is usually obtained by multi-reflection between an anode and a cathode. Since top emitting OLED devices often use a reflective anode and a semi-transparent cathode, micro-cavitation necessarily occurs. Therefore, the understanding of micro-cavitation is important in the design of OLED devices. An example of a top emitting micro-cavity structure is shown in Fig. 5.3.

Since the micro-cavity structure provides changes in emission spectra, emission profile, efficiency, etc., it can be utilized for the improvement of OLED device performance.

The micro-cavity effect can be analyzed as a kind of optical filter that selects and emphasizes the internal emission from the organic layer. The micro-cavity condition is represented by the following equation [31]:

$$2L/\lambda_{max} + \Phi/(2\pi) = m\,(m:integer)$$

where L is the optical path length between two reflective layers, λ_{max} is the peak wavelength of the micro-cavity, and Φ is the sum of the phase shift from the reflection at the anode and the cathode.

The full width at half maximum (FWHM) can be estimated by the following equation, where R_1 and R_2 are the reflectances of the two mirrors [31–33]:

$$\text{FWHM} = \frac{\lambda \text{max}^2}{2L} \times \frac{1 - \left(R_1 R_2\right)^{1/2}}{\pi \left(R_1 R_2\right)^{1/4}}$$

In micro-cavity OLEDs, the shape of the emission spectrum can be controlled in accordance with optical conditions, and thus the emission intensity in the normal direction can be enhanced. Therefore, micro-cavity technologies contribute to high color saturation, high efficiency, etc.

According to the equation above, the peak wavelength is dependent on the optical path length. Therefore the optical path length should be designed for each color – red, green, and blue.

Sony developed and commercialized an AM-OLED-TV by using micro-cavity structure [5, 33, 34]. The specifications of their 13″ AM-OLED prototype display are given in Table 5.2, which shows that high color purities of three colors were obtained as a result of the micro-cavity effect.

Hsu et al. of National Chiao Tung University, etc. (Taiwan, ROC) reported on the experimental results of the relationship between the thickness of ITO layer and the peak wavelength of the micro-cavity [35]. Based on their report, the relationship between the optical path length between two reflective layers and the peak wavelength of the micro-cavity is shown in Fig. 5.18.

Micro-cavity effects in top emitting white OLED devices have also been investigated [29, 36, 37]. For applying top emitting white OLED devices to full-color displays, the optical path lengths for red, green, and blue pixels should be adjusted and optimized, respectively, as shown in Fig. 5.19.

In addition, in bottom emitting OLEDs, the micro-cavity effect can be utilized. For example, Kim et al. of Kookmin University, etc. (Republic of Korea) obtained a deep blue bottom

Table 5.2 Specifications of 13.0″ AM-OLED prototype display developed by Sony [5]

Size	13.0″ diagonal
Number of pixels	800RGB×600 (SVGA)
Pixel pitch	330 μm×330 μm
Color coordinate (CIE)	R (0.66, 0.34)
	G (0.26, 0.65)
	B (0.16, 0.06)
Color temperature (white)	9300 K
Peak luminance	>300 cd/m²
Contrast ratio	1400:1 (in dark)
	200:1 (under 500 lx)

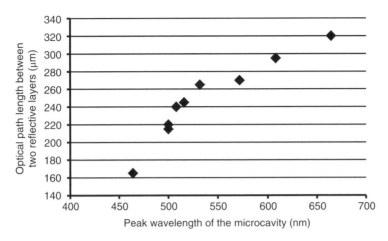

Figure 5.18 The relationship between the optical path length between two reflective layers and the peak wavelength of the micro-cavity (data source [35])

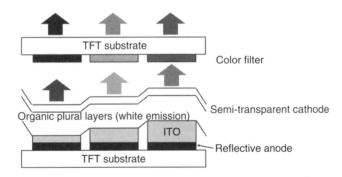

Figure 5.19 An example of top emitting OLED display with white emission combined with micro-cavity structure [29]

emitting OLED devices with excellent color coordinate (e.g. $x = 0.139$. $y = 0.081$) by applying a Bragg mirror layer between an ITO anode and a glass substrate [38]. The Bragg mirror consists of a SiO_2/TiO_2 multi-layer.

5.6 Multi-Photon OLED

Multi-photon technology is an important technology in OLEDs because it can give high efficiency and long lifetime. In particular, it is very useful for OLED lighting.

Multi-photon technology was first reported by Kido and Matsumoto et al. of Yamagata University and International Manufacturing and Engineering Services (IMES) (Japan) [39–41]. The multi-photon OLED structure comprises multiple emissive units which are connected by so-called charge generation layers (CGLs) or transparent conducting materials such as ITO

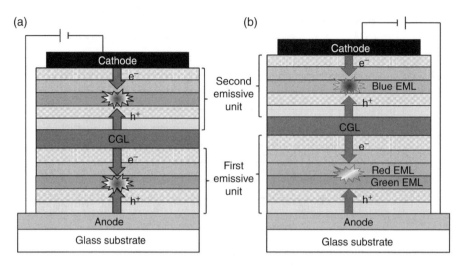

Figure 5.20 Schematic views of a multi-photon OLED device with a charge generation layer. (a) Same emissive units connected by the charge generation layer. (b) Different emissive units connected by the charge generation layer

Table 5.3 Examples of the effect of multi-photon OLED [42, 43]

Number of stacking	1	2	3
Driving of 2.0 mA/cm²	1600 cd/m² 3.26 V	3100 cd/m² 7.19 V	4300 cd/m² 10.35 V
Luminance of 100 cd/m²	90 cd/A	181 cd/A	236 cd/A
Luminance of 1000 cd/m²	83 cd/A	171 cd/A	244 cd/A

[43]. OLED devices in which multiple emissive units are connected by conducting materials are often called tandem structure. On the other hand, in the case of multi-photon OLED devices with CGLs, the charge generation material is not required to be conductive but has an insulating or semiconducting property. At the charge generation layer, positive and negative charges are generated and injected into the adjacent emissive units. The schematic view of the example is shown in Fig. 5.20, which shows two stacked multi-photon OLED, but further stacking such as three, four, or more stacked units is possible. The stacked units must not be the same as each other, although Fig. 5.20(a) shows a stacking example with the same *emissive* layers. In Fig. 5.20(b), different emissive units can be connected to the charge generation layer. Due to the charge generation phenomenon, current efficiency (cd/A) of OLED devices is almost linearly correlated with the stacking number. Therefore, lifetime at a given luminance is drastically improved by increasing the stacking number, because lifetime is related to current density. However, the driving voltage increases with increased stacking number. Therefore, the effect of multi-photon structure on power efficiency (lm/W) is limited.

Table 5.3 shows an example of performance obtained in multi-photon emission OLEDs with the green phosphorescent materials shown in Fig. 5.21 [42, 43]. Under the driving current of 2.0 mA/cm², the one-unit device shows a luminance of 1600 cd/m² at a voltage of 3.26 V.

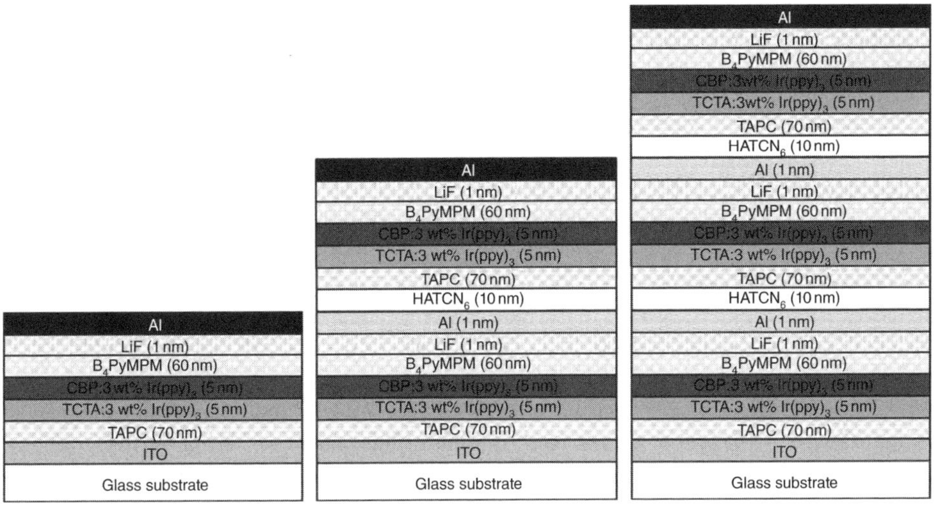

Figure 5.21 Actual example of multi-photon emission OLEDs [43]. "Copyright 2011 The Japan Society of Applied Physics"

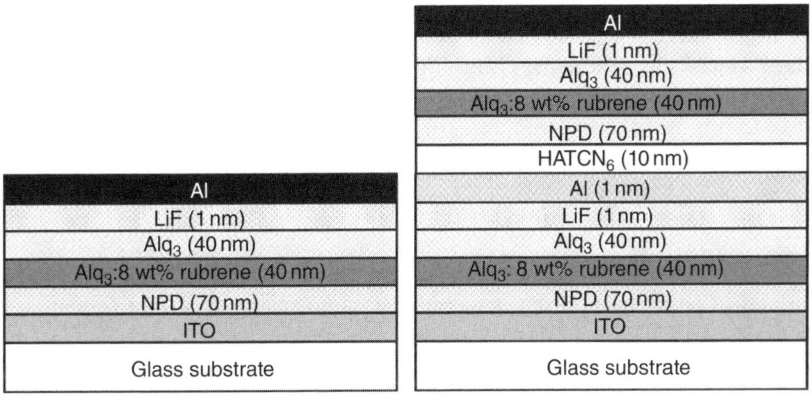

Figure 5.22 Actual example of multi-photon emission OLEDs [43]. "Copyright 2011 The Japan Society of Applied Physics"

At the same driving current, the two-unit and three-unit devices show $3100\,cd/m^2$ at $7.19\,V$ and $4300\,cd/m^2$ at $10.35\,V$, respectively. The luminance almost proportionally increases with increasing staking number, while the voltage also increases with the staking number. As is obvious from Table 5.3, the current efficiencies increase with increasing staking number at luminances of $100\,cd/m^2$ and $1000\,cd/m^2$, respectively.

Another example is shown in Fig. 5.22 and Table 5.4 [43]. Under the driving current of $30\,mA/cm^2$, the one-unit device shows a luminance of $2240\,cd/m^2$ at $7.9\,V$. At the same driving current, the two-unit device shows $4111\,cd/m^2$ at $13.4\,V$. It is found that the luminance of the two-unit device is almost twice that of the one-unit device, while the voltage of the two-unit device is also almost twice. To obtain $1000\,cd/m^2$, the one-unit device would require $12.0\,mA/cm^2$

Table 5.4 Examples of the effect of multi-photon OLED [43]

Number of stacking	1	2
Driving of 30 mA/cm²	2240 cd/m²	4111 cd/m²
	7.9 V	13.4 V
Luminance of 1000 cd/m²	12.0 mA/cm²	6.04 mA/cm²
Lifetime evaluation under	T_{85} = 98H	T_{85} = 190H
50 mA/cm²	(initial luminance = 3410 cd/m²)	(initial luminance = 6600 cd/m²)

but the two-unit device requires only 6.04 mA/cm². This result indicates that the current efficiency of the two-unit device is about twice of that of the one-unit device. In addition, the two-unit device shows longer lifetime, due to the lower current density. At a driving current of 50 mA/cm², the lifetime of T_{85} (driving period giving 85% of the initial luminance) of the one-unit device is 98 hours. In this case, the initial luminance is 3410 cd/m². On the other hand, the two-unit device has double the lifetime T_{85} at 190 hours under the same driving current, although the initial luminance is 6600 cd/m², which is almost twice that of the one-unit device.

The multi-photon structure is also useful for obtaining white emission by stacking emission layers of different colors, as shown in Fig. 5.20(b).

Charge generation layers basically consist of a combination of an n-type layer and a p-type layer. The n-type layer is typically composed of electron injecting metal compounds such as LiF, or metal-doped electron transporting organic layers doped with metals such as Li [44], Cs, or Mg [45]. The p-type layer is typically composed of metal oxides such as indium tin oxide (ITO) [40] V_2O_5 [42], or MoO_3 [46]. or p-doped hole transporting organic layers doped with $FeCl_3$ [44], tetrafluorotetracyano-quinodimethane (F_4-TCNQ) [45], etc.

5.7 Encapsulation

When bare OLED devices are used in an ambient atmosphere, dark spot formation leads to complete device degradation within a few hours. This phenomenon is attributed to the attack of water and oxygen in the ambient air. Therefore, encapsulating technology is very important.

One typical and simple device structure is shown in Fig. 5.23. In this structure, an OLED device is sandwiched by a counter substrate, and the sandwiched structure is sealed by using a resin such as epoxy. The typical pattern sizes of the resin are 10 μm in height and 1 mm in width. The space between the OLED substrate and the counter substrate is filled with dry N_2. In 1994, Burrows et al. of Forrest's group at Princeton University reported that the OLED device with a similar structure to that shown in Fig. 5.23 showed a greatly enhanced lifetime compared with the reference OLED with no encapsulation [47].

The degradation induced by attack from the ambient air has been studied from the early stage of the research and development of OLEDs.

Savvate'ev et al. of the Hebrew University, etc. (Israel) reported the degradation phenomena of polymer OLED devices with no encapsulation [48].

Schaer and coworkers at the Laboratoire de Physique des Solides Semi-cristallins (Switzerland) and CFG Microelectronic (Switzerland) investigated water vapor and oxygen

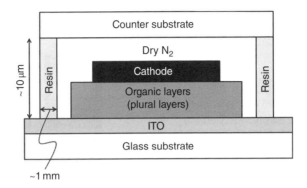

Figure 5.23 An example of encapsulating OLED

degradation mechanisms in OLED devices [49]. According to their report, no dark spot and no significant degradation appears when OLED devices are operated for 2500 hours under vacuum at a pressure lower than 10^{-5} mbar or pure inert gas atmosphere, in which the residual water vapor and residual contaminants are less than 1 ppm. On the other hand, when the OLED devices are operated under a relative humidity of 75% and applied bias at room temperature, the device degradation is reported to be very fast, leading to complete device failure within a few hours.

The relationship between the degradation and the encapsulating technologies are explained by using three typical examples.

Figure 5.24 shows the first example where an OLED device is seriously damaged under storage due to the poor encapsulation. This OLED device is encapsulated by a PEN film with a SiNx passivation layer possessing poor encapsulating ability. The surrounding of the device is covered by a UV-resin sealant with the thickness of about 10 μm. After storage for a certain time at conditions of 60°C/90%RH, the emission of OLED devices was observed. Even after storage for 2 hours, a growth of dark spots and emission shrinkage was observed. As shown in Fig. 5.24, the emission shrinkage occurs at the position of the edge of the Al cathode, due to the penetration of H_2O and O_2, because the Al cathode has a certain protecting ability against H_2O and O_2. For quantitative evaluations, the increase in such shrinkage seems to be more useful than the increase in the size and number of dark spots. After storage for 20 hours, the growth of the emission shrinkage and dark spots is serious. In this case, the attack of ambient air through the PEN film is more severe than through the surrounding UV-resin.

As the second example, Fig. 5.25 shows the degradation of the OLED device encapsulated by a counter glass substrate. In this case, the glass substrate can be considered to have a perfect protection against the ambient air. However, the ambient water and oxygen penetrates the OLED device through the bulk of the UV resin and at the interface between the substrate and the UV resin. In this experiment, the width and thickness of the UV resin is about 2 mm and 10 μm, respectively. In a storage time of less than about 100 hours under conditions of 60°C/90%RH, no serious damage occurs, while there is a slight growth in the size of dark spots. However, when the storage time is longer than about 100 hours, the obvious growth in the emission shrinkage and in the dark spot occurs.

The third example is shown in Fig. 5.26. In this case, a desiccant is added to the second case. Here, no damage is observed, being different from Figs 5.24 and 5.25, because the penetrated moisture and oxygen is trapped by the desiccant before it is able to attack the OLED device.

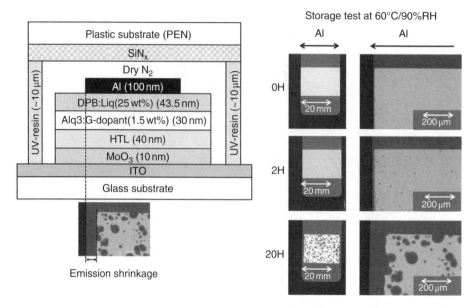

Figure 5.24 An example of degradation in OLED devices with poor encapsulation

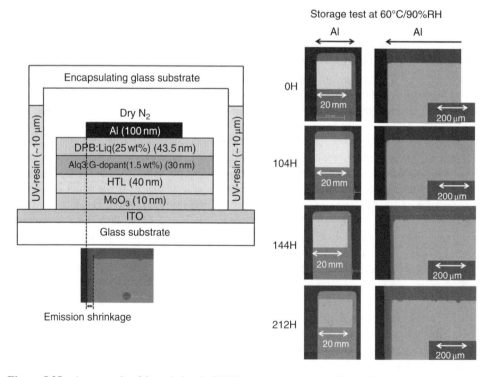

Figure 5.25 An example of degradation in OLED devices encapsulated by a glass substrate. The width of UV resin is 2 mm

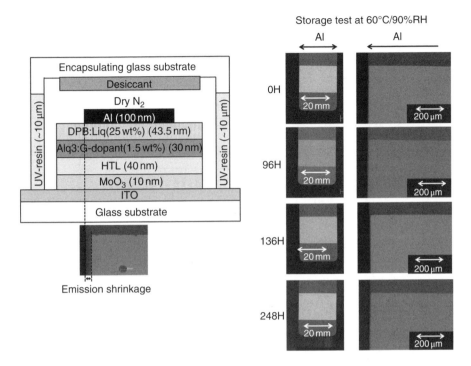

Figure 5.26 An example of degradation in OLED devices encapsulated by a glass substrate and a desiccant. The width of UV resin is 2 mm

Figure 5.27 shows the relationship between the growth of emission shrinkage and storage time in the typical three encapsulating conditions described in Figs 5.24, 5.25, and 5.26.

In the second case, it is obvious that the growth of emission shrinkage is remarkably better than the first case with poor encapsulation. In addition, after 100 hours, the vigorous growth of emission shrinkage starts and the growth is almost proportional to time. In other words, the encapsulating ability of the surrounding UV-resin is restricted.

Therefore, many commercial OLED products utilize desiccants, while other technologies without desiccant are also known.

Three typical practical encapsulating structures are usually used in commercial OLEDs, as illustrated in Fig. 5.28.

The first is the use of a desiccant, as shown in Fig. 5.28(a). In this method, OLED devices are encapsulated by a counter substrate, being similar to liquid crystal devices. The edge of the device is sealed by UV-cured epoxy resin. The resin is required to give no damage to OLED performance. The thickness of the sealant is about $5 \sim 10\,\mu m$. A desiccant material is set in the encapsulated OLED devices. Since the thickness of the commonly used desiccant is $200 \sim 300\,\mu m$, the counter substrate tends to a non-flat shape, as shown in Fig. 5.28(a). As the counter substrate, glass or metal are often used. Typical materials for the desiccant are CaO and BaO.

The second is the frit glass seal method, as shown in Fig. 5.28(b). The frit glass paste is put between a glass substrate with OLED device and a counter glass substrate. Using irradiation by a laser of part of the frit glass paste, the paste melts by heat, and is combined with both substrates. The penetration of moisture and oxygen is avoided by this glass seal, and high

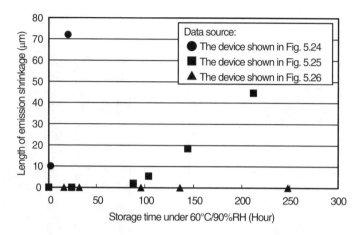

Figure 5.27 The relationship between growth of the emission shrinkage and storage time in the typical three encapsulating conditions described in Figs 5.24, 5.25, and 5.26. The width of UV resin is 2 mm

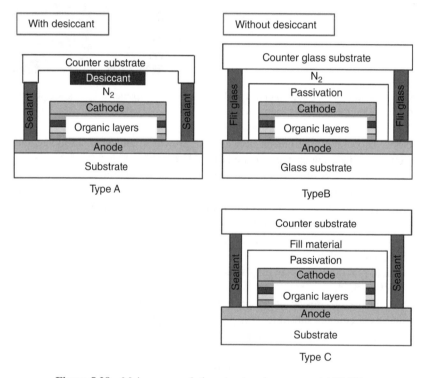

Figure 5.28 Major encapsulating structure in commercial OLEDs

reliability is obtained. However, this method has some limitations due to the heat generation in the laser process. For example, this is not easily applied to plastic films.

The third is the combination of thin film passivation and counter substrate, as shown in Fig. 5.28(c). In this method, OLED parts including organic materials are encapsulated by a thin film deposited by CVD, sputtering, etc., and then encapsulated by a counter substrate.

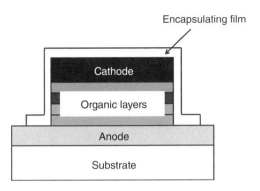

Figure 5.29 Thin film encapsulating technology

While the above-mentioned two methods provides a gas area in the OLED devices, this third method can produce a complete solid OLED device. The merit of this third method is it being an all solid device, but the problem is the high cost for fabrication of thin film deposition. One thin film layer does not tend to achieve sufficient reliability, so multiple layers and/or thick film with μm order is required.

5.7.1 Thin Film Encapsulation

Thin film encapsulating technologies seem to be an ideal and simple method, as shown in Fig. 5.29. Several technologies and reports are reviewed below.

Kim et al. of ELiA Tech Co. Ltd (South Korea) achieved almost no degradation for 500 hours under storage conditions of 65 °C/99%RH in OLEDs encapsulated by thin film multi-layer [50]. In their thin film multi-layer, they deposited an organic protection layer, first polymer layer, first inorganic layer, second inorganic layer and second polymer layer with this sequence on OLED devices. The polymer layers were deposited by screen printing, followed by UV exposure. The inorganic layers were deposited by PE-CVD.

Catalytic chemical vapor deposition (Cat-CVD) technology has been developed and applied to OLEDs [51]. Cat-CVD can deposit SiN_x, SiO_xN_y, etc. at 100 °C without thermal damage to the OLED devices. Ogawa et al. of Japan Advanced Institute of Science and Technology (JAIST), etc. (Japan) deposited a bi-layer structure of 200 nm thick SiN_x and 300 nm thick SiO_xN_y on a PET film, obtaining a water vapor transmission rate (WVTR) of 0.018 g/m²/day [51]. This WVTR value is much lower than that of the PET film with no gas barrier film (4.6 g/m²/day). They also applied the 300 nm thick stacked film composed of SiN_x and SiO_xN_y to a passivation layer of OLED, observing no damage after 1000 hours under storage conditions of 60 °C/90%RH. They commented that the 1000 hours corresponds to 50,000 hours under normal temperature and humidity conditions, assuming an acceleration factor of 50.

Li et al. of Holst Centre/TNO and Philips Research Laboratories (Netherlands) developed a thin film silicon nitride (SiNx) barrier stack, fabricated by low-temperature plasma enhanced chemical vapor deposition (PE-CVD) [52]. They reported that the SiNx film has an intrinsic moisture barrier of WVTR < 10^{-6} g/m²/day. By applying the SiNx film on OLED devices, no growth of dark spots was observed in accelerated storage lifetime test for 500 hours at 60°C/90%RH.

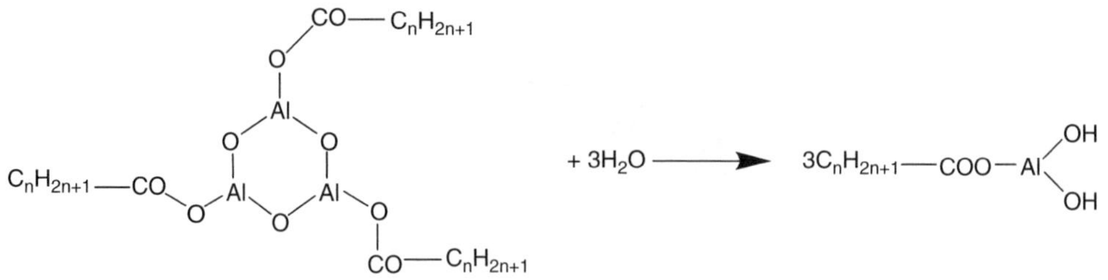

Figure 5.30   General formula for OleDry and its reaction with moisture [53, 54]

5.7.2   Desiccant Technologies

While classical desiccants for OLED devices are inorganic materials, which are physical adsorption materials such as silica gels and molecular sheaves, and chemical reaction materials such as BaO and CaO, new desiccant materials have also been developed.

Futaba Corporation developed a transparent liquid desiccant named OleDry [53, 54]. The desiccant is mainly composed of organic metal compound, as shown in Fig. 5.30. This material can be applied to Type C in Fig. 5.28. They reported that the material is a transparent liquid with the higher transmittance than 98% (film thickness: 12 μm) and therefore can be applicable to not only bottom emission OLEDs but also top emission OLEDs. Since the compounds in OleDry chemically react with moisture, no re-emission of moisture occurs due to heat, in contrast to silica gel and molecular sheaves.

JSR Corporation (Japan) reported a coatable and/or printable transparent desiccant that can be transferred to solid phase [55,56]. The desiccant consists of organic metal compounds and binder polymers. The desiccant is transparent and is a non-solvent type coating material. The materials can be put on a substrate by the one drop fill (ODF) method, giving the OLED device structure of Type C in Fig. 5.28. When the desiccant was applied to an OLED device with a 1 μm thick SiNx passivation layer deposited by surface-wave plasma CVD (SWP-CVD), the OLED device was reported to show no defect in a 60°C/90%RH test for 500 hours, while the reference OLED without the desiccant was reported to show some dark spots in the same storage reliability test.

References

[1] D. R. Baigent, R. N. Marks, N. C. Greenham, R. H. Friend, S. C. Moratti and A. B. Holmes, *Appl. Phys. Lett.*, **65**, 2636–2638 (1994).

[2] C. W. Chen, P. Y. Hsieh, H. H. Chiang, C. L. Lin, H. M. Wu, C. C. Wu, *Appl. Phys. Lett.*, **83(25)**, 5127–5129 (2003).

[3] S. F. Hsu, C.-C. Lee, A. T. Hu, C. H. Chen, *Current Applied Physics*, **4**, 663–666 (2004); S.-F. Hsu, C.-C. Lee, S.-W. Hwang, H.-H. Chen, C. H. Chen, A. T. Hu, *Thin Solid Films*, **478**, 271–274 (2005).

[4] L. S. Hung, C. W. Tang, M. G. Mason, P. Raychaudhuri, J. Madathil, *Appl. Phys. Lett.*, **78(4)**, 544–546 (2001).

[5] T. Sasaoka, M. Sekiya, A. Yumoto, J. Yamada, T. Hirano, Y. Iwase, T. Yamada, T. Ishibashi, T. Mori, M. Asano, S. Tamura, T. Urabe, *SID 01 Digest*, 24.4 L(p. 384) (2001).

[6] V. Bulovic, G. Gu, P. E. Burrows, S. R. Forrest, M. E. Thompson, *Nature*, **380**, 29–29 (1996); G. Gu, V. Boluvic, P. E. Burrows, S. R. Forrest and M. E. Thompson, *Appl. Phys. Lett.*, **68**, 2606–2608 (1996).

[7] K. Morii, M. Ishida, T. Takashima, T. Shimoda, Q. Wang, M. K. Nazeeruddin, M. Grätzel, *Appl. Phys. Lett.*, **89**, 183510 (2006).

[8] Y. Meng, W. Xie, N. Zhang, S. Chen, J. Li, W. Hu, Y. Zhao, J. Hou, S. Liu, *Microelectronics Journal*, **39**, 723–726 (2008).

[9] K.-H. Kim, S.-Y. Huh, S.-M. Seo, H. H. Lee, *Organic Electronics*, **9**, 1118–1121 (2008).

[10] J. K. Noh, M. S. Kang, J. S. Kim, J. H. Lee, Y. H. Ham, J. B. Kim, M. K. Joo, S. Son, *Proc. IDW'08*, OLED3–1 (p. 161) (2008).

[11] Y. Lee, J. Kim, S. Kwon, C.-K. Min, Y. Yi, J. W. Kim, B. Koo, M. P. Hong, *Organic Electronics*, **9**, 407–412 (2008).

[12] H. Fukagawa, K. Morii, M. Hasegawa, Y. Nakajima, T. Takei, G. Motomura, H. Tsuji, M. Nakata, Y. Fujisaki, T. Shimizu, T. Yamamoto, *SID 2014 Digest*, P-154 (p. 1561) (2014).

[13] J. Kido, K. Hongawa, K. Okuyama and K. Nagai, *Appl. Phys. Lett.*, **64(7)**, 815–817 (1994).

[14] J. Kido, M. Kimura and K. Nagai, *Science*, **267**, 1332–1334 (1995).

[15] Y. Sun, N. C. Giebink, H. Kanno, B. Ma, M. E. Thompson, S. R. Forrest, *Nature*, **440**, 908–912 (2006).

[16] S. Tokito, T. Iijima, T. Tsuzuki, F. Sato, *Appl. Phys. Lett.*, **83(3)**, 569–571 (2003).

[17] S. Tokito, T. Tsuzuki, F. Sato, T. Iijima, *Current Appl. Phys.*, **5**, 331–336 (2005).

[18] G. Cheng, Y. Zhang, Y. Zhao, Y. Lin, C. Ruan, S. Liu, T. Fei, Y. Ma, Y. Cheng, *Appl. Phys. Lett.*, **89**, 043504 (2006).

[19] S. Reineke, F. Lindner, G. Schwartz, N. Seidler, K. Walzer, B. Lüssem, K. Leo, *Nature*, **459**, 234 (2009).

[20] B. W. D'Andrade, R. J. Holmes, S. R. Forrest, *Adv. Mater.*, **16(7)**, 624–628 (2004).

[21] K. Mameno, R. Nishikawa, T. Omura, S. Matsumoto, S. A. VanSlyke, A. D. Arnold, T. K. Hatwar, M. V. Hettel, M. E. Miller, M. J. Murdoch, J. P. Spindler, *Proc. IDW'04*, AMD2/OLED4–1 (2004).

[22] C. Hosokawa, M. Eida, M. Matsuura, K. Fukuoka, H. Nakamura, T. Kusumoto, *Synth. Met.*, **91**, 3–7 (1997).

[23] C. Hosokawa, E. Eida, M. Matsuura, F. Fukuoka, H. Nakamura, T. Kusumoto, *SID 97 Digest*, L2.3, p. 1037 (1997).

[24] C. Hosokawa, M. Eida, M. Matsuura, H. Nakamura, T. Kusumoto, *J. SID*, **5**, 331 (1997).

[25] M. Matsuura, M. Eida, M.Funahashi, K. Fukuoka, H. Tokairin, C. Hosokawa, T. Kusumoto, *Proc. IDW'97*, **581** (1997).

[26] P. E. Burrows, V. Khal"n, G. Gu, S. R. Forrest, *Appl. Phys. Lett.* **73**, 435 (1998).

[27] J. Kido, Y. Yamagata, G. Harada, *Extended Abstracts of the 44th Spring Meeting 1997, Japan Society of Applied Physics and Related Societies*, 29-NK-14, p. 1156 (1997).

[28] K. Mameno, S. Matsumoto, R. Nishikawa, T. Sasatani, K. Suzuki, T. Yamaguchi, K. Yoneda, Y. Hamada, N. Saito, *Proc. IDW'03*, AMD4/OEL5-1 (p. 267) (2003).

[29] M. Kashiwabara, K. Hanawa, R. Asaki, I. Kobori, R. Matsuura, H. Yamada, T. Yamamoto, A. Ozawa, Y. Sato, S. Terada, J. Yamada, T. Sasaoka, S. Tamura, T. Urabe, *SID 04 Digest*, 29.5L (p. 1017) (2004).

[30] T. Tsujimura, S. Mizukoshi, N. Mori, K. Miwa, Y. Maekawa, M. Kohno, K. Onomura, K. Mameno, T. Anjiki, A. Kawakami, S. VanSlyke, *Proc. IDW'08*, OLED2-1 (p. 145) (2008).

[31] M. Kashiwabara, K. Hanawa, R. Asaki, I. Kobori, R. Matsuura, H. Yamada, T. Yamamoto, A. Ozawa, Y. Sato, S. Terada, J. Yamada, T. Sasaoka, S. Tamura and T. Urabe, *SID 04 Digest*, 29.5L(p. 1017) (2004).

[32] E. F. Schubert, N. E. J. Hunt, M. Micovic, R. J. Malik, D. L. Sivco, A. Y. Cho, G. J. Zydzik, *Science*, **256**, 943–945 (1994).

[33] J. Yamada, T. Hirano, Y. Iwase, T. Sasaoka, Proc. *AM-LCD'02*, OD-2 (p. 77) (2002).

[34] T. Urabe, *Proc. IDW'03*, AMD3/OEL4-1 (p. 251) (2003).

[35] S. F. Hsu, C.-C. Lee, A. T. Hu, C. H. Chen, *Current Applied Physics*, **4**, 663–666 (2004).

[36] A. Chen, H.-S. Kwok, *Organic Electronics*, **12**, 2065–2070 (2011).

[37] J. Cao, X. Liu, M. A. Khan, W. Q. Zhu, X. Y. Jiang, Z. L. Zhang, S. H. Xu, *Current Applied Physics*, **7**, 300–304 (2007).

[38] H. K. Kim, S.-H. Cho, J. R. Oh, Y.-H. Lee, J.-H. Lee, J.-G. Lee, S.-K. Kim, Y.-I. Park, J.-W. Park, Y. R. Do, *Organic Electronics*, **11**, 137–145 (2010).

[39] J. Kido, J.Endo, T.Nakada, K.Mori, A.Yokoi, T.Matsumoto, *Japan Society of Applied Physics, 49th Spring Meet.*, Ext.Abstr(p. 1308), 27p-YL-3 (2002).

[40] T. Matsumoto, T. Nakada, J. Endo, K. Mori, N. Kawamura, A. Yokoi, J. Kido, *SID 03 Digest*, 27.5L (p. 979) (2003).

[41] T. Matsumoto, T. Nakada, J. Endo, K. Mori, N. Kawamura, A. Yokoi, J. Kido, *Proc. IDW'03*, OEL2-1 (2003).

[42] J. Kido, H. Sasabe, D. Yokoyama, Y. J. Pu, *SID 2012 Digest*, 57.2 (2012).

[43] Y. J. Pu and J. Kido, *Oyobutsuri*, **80(4)**, 295–299 (2011).

[44] L. S. Liao, K. P. Klubek, C. W. Tang, *Appl. Phys. Lett.*, **84(2)**, 167–169 (2004).

[45] C. W. Law, K. M. Lau, M. K. Fung, M. Y. Chan, F. L. Wong, C. S. Lee, S. T. Lee, *Appl. Phys. Lett.*, **89**, 133511 (2006).

[46] H. Kanno, R. J. Holmes, Y. Sun, S. Kena-Cohen, S. R. Forrest, *Adv. Mater.*, **18**, 339–342 (2006).

[47] P. E. Burrows, B. Bulovic, S. R. Forrest, L. S. Sapochak, D. M. McCarty, M. E. Thompson, *Appl. Phys. Lett.*, **65(23)**, 2922–2924 (1994).

[48] V. N. Savvate'ev, A. V. Yakimov, D. Davidov, R. M. Pogreb, R. Neumann, Y. Avny, *Appl. Phys. Lett.*, **71(23)**, 3344–3346 (1997).

[49] M. Schaer, F. Nüesch, D. Berner, W. Leo, L. Zuppiroli, *Adv. Funct. Mater.*, **11(2)**, 116–121 (2001).

[50] K. M. Kim, B. J. Jang, W. S. Cho, S. H. Ju, *Current Applied Physics*, **5**, 64–66 (2005).

[51] Y. Ogawa, K. Ohdaira, T. Oyaidu, H. Matsumura, *Thin Solid Films*, **516**, 611–614 (2008); A. Heya, T. Minamikawa, T. Niki, S. Minami, A. Masuda, H. Umemoto, N. Matsuo, H. Matsumura, *Thin Solid Films*, **516**, 553–557 (2008).

[52] F. M. Li, S. Unnikrishnan, P. van de Weijer, F. van Assche, J. Shen, T. Ellis, W. Manders, H. Akkerman, P. Bouten, T. van Mol. *SID 2013 Digest*, 18.3 (p. 199) (2013).

[53] Y. Tsuruoka, S. Hieda, S. Tanaka, H. Takahashi, *SID'03 Digest*, 21.2 (p. 860) (2003).

[54] T. Niiyama, S. Tanaka, Y. Hoshina, M. Sisikura, and R. Kajiyama, *SID 2013 Digest*, 55.3(2013); Y. Hoshina, T. Niyama, S. Tanaka, M. Miyagawa, *Proc. IDW'13*, OLED4–5L (2013).

[55] K. Konno, T. Arai, M. Takahashi, T. Kajita, *13th Japanese OLED Symposium*, S3–2 (2011); T. Arai, K. Konno, T. Miyasako, M. Takahashi, M, Nishikawa, K. Azuma, T. Ueno, and M. Hasuta, *15th Japanese OLED Symposium*, S7–4 (2012).

[56] H. Katsui, T. Miyasako, T. Arai, M. Takahashi, N. Onimaru, N. Takamatsu, T. Yamamura, K. Konno, K. Kuriyama, *Proc. IDW'14*, OLED3–3 (2014).

6

OLED Fabrication Process

Summary

This chapter describes the fabrication processes of OLED devices. In general, OLEDs are fabricated by either a dry process or a wet process. The commonly used dry process is vacuum evaporation. Three types of vacuum evaporation techniques are reviewed. In wet processes, there are processes for coating without fine patterning and processes for making fine patterns.

In addition to deposition techniques of each organic layer, RGB patterning technologies are important. These RGB patterning technologies are vacuum deposition using fine metal masks, wet processes such as ink-jet, nozzle printing, relief printing, etc., and laser patterning process.

Key words

Process, dry process, wet process, vacuum deposition, mask deposition, solution, patterning, coating, laser, ink-jet, nozzle printing, relief printing

6.1 Vacuum Evaporation Process

The vacuum evaporation process is the most common technology for fabricating OLED devices. The schematic view of this method is shown in Fig. 6.1. In this process, organic materials set into crucibles are changed to gas phase by elevating their temperature at low pressures such as $10^{-5} \sim 10^{-7}$ torr. The materials in gas phase are deposited on substrates, being changed to solid phase because the temperature of the substrate is much lower than the crucible.

OLED Displays and Lighting, First Edition. Mitsuhiro Koden.
© 2017 John Wiley & Sons, Ltd. Published 2017 by John Wiley & Sons, Ltd.

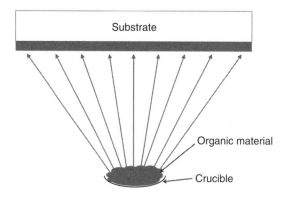

Figure 6.1   A schematic view of the vacuum evaporation method

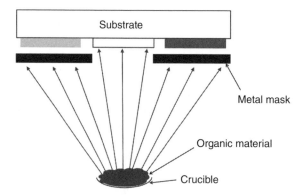

Figure 6.2   The schematic view of mask deposition

6.1.1   Mask Deposition

In order to fabricate sub-pixel emitting patterns for realizing full color, shadow mask pattern-ing processes are widely used [1]. The schematic view of mask deposition is illustrated in Fig. 6.2. While shadow mask deposition is a simple technology in principle, actual shadow mask deposition is not so easy, especially for high resolution displays and large size displays because of mask distortion or the high risk of defects.

6.1.2   Three Types of Evaporation Methods

From the relationship between the evaporation source and the substrate, three types of evapo-ration technologies are known: point source, linear source and planar source evaporation. These methods are shown in Fig. 6.3.

Point source is the simplest method and widely used in R&D and the production of small or medium size substrates. In this method, an evaporation material is set into the point type of crucible (source) and a substrate is set at a certain distance from the crucible. The rotation of the substrate is often used for obtaining good thickness uniformity of deposited organic layers.

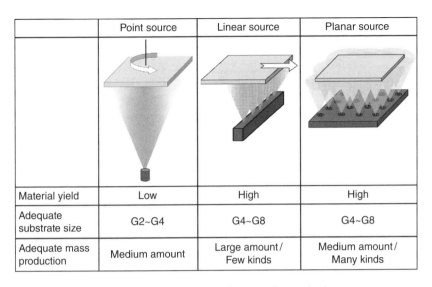

	Point source	Linear source	Planar source
Material yield	Low	High	High
Adequate substrate size	G2~G4	G4~G8	G4~G8
Adequate mass production	Medium amount	Large amount / Few kinds	Medium amount / Many kinds

Figure 6.3 Three types of evaporation methods

In addition, in order to obtain good thickness uniformity of deposited organic layers and in order to reduce the influence of the radiant rays to the substrate, the distance between the target (source) and the substrate (the T/S distance) is very large. The distance is usually several tens of cm. Due to these large T/S distances, most evaporated materials are not deposited on the substrate but on the walls of the vacuum chamber. Therefore, the material yield of point source evaporation is usually less than 10%. If this method is applied to large size substrates, the evaporation equipment would have to be extremely large. Therefore, the use of point source evaporation is restricted to small and medium size substrates. The upper limit of the substrate size seems to be G4 (730×920 mm).

By contrast, the linear source method is applicable to large size substrates. In this method, a linear shape evaporation source is used and the substrate moves. The T/S distance can be shorter than the point source method. Therefore, it is possible to obtain high material yield.

Another method is planar source evaporation, in which the evaporation source has a planar shape and the substrate does not move. The T/S distance can be shortened as it can in the linear source method. Fujimoto et al. of Hitachi Zosen Corporation developed planar source evaporation equipment, as shown in Fig. 6.4 [2]. They simulated the material yield evaporated by the planar source, assuming that the T/S distance is 200 mm and thickness uniformity is better than ±3%. The simulated results are shown in Fig. 6.5. When the substrate size is not so large, such as G2 or G3, the simulated material yield is about 20–30%. By contrast, when the substrate size is as large as G6–G8, the material yield is higher than 60–70%. Indeed, they have fabricated a planar source evaporation system for G6 substrate.

6.1.3 Ultra-High Vacuum

In vacuum deposition, it has been reported that the pressure and residual water play a significant role in OLED performance.

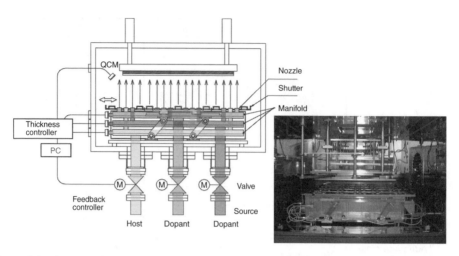

Figure 6.4 An example of planer source evaporation [2]. (Source: supplied by Mr Y. Matsumoto of Hitach Zosen Corporation)

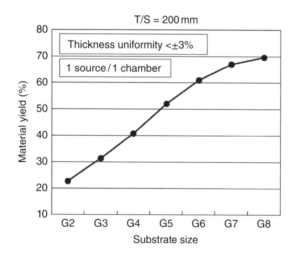

Figure 6.5 Simulated material yield in planar source evaporation [2]. (Source: supplied by Mr Y. Matsumoto of Hitach Zosen Corporation)

Ikeda et al. of Japan Advanced Institute of Science and Technology (JAIST), PRESTO and Kitano Seiki investigated the effect of pressure in the vacuum chamber on OLED performance [3]. While the vacuum pressure in depositions of OLED materials is usually range of 10^{-5} to 10^{-7} torr, they prepared a deposition chamber having a pressure range of 10^{-7} to 10^{-9} torr. They fabricated OLED devices with the structure of glass/ITO(150 nm)/CuPc(10 nm)/ α-NPD(50 nm)/Alq$_3$(65 nm)/LiF(5 nm)/Al(80 nm) under three different vacuum conditions. Table 6.1 shows the relationship between the vacuum pressure and the residual gas components observed during the deposition. They reported that the dominant gas components during device fabrication were water for device A, water/nitrogen for device B, and nitrogen for

Table 6.1 The influence of vacuum pressure on the residual gas components observed during the deposition and lifetime performance [3]

Device	Pressure (torr)		Ion current (A)				Half lifetime LT_{50} [*] (Hours)
	Base	Process	H_2O^+ (m/z = 18)	N_2^+ or CO^+ (m/z = 28)	O_2^+ (m/z = 32)	CO_2^+ (m/z = 44)	
A	5.0×10^{-7}	3.2×10^{-7}	3.8×10^{-8}	2.0×10^{-9}	4.0×10^{-10}	5.5×10^{-10}	1.2
B	4.0×10^{-8}	4.8×10^{-8}	2.7×10^{-9}	1.6×10^{-9}	3.3×10^{-10}	4.1×10^{-10}	24.2
C	2.0×10^{-9}	1.6×10^{-8}	5.9×10^{-10}	1.0×10^{-9}	6.9×10^{-11}	6.9×10^{-10}	31.0

* Under a constant dc current drive with 250 mA/cm^2, which gives initial luminance (L_0) of about 10,000 cd/m^2

device C. Table 1 also showed the half lifetime (LT_{50}) under a constant dc current of 250 mA/cm^2, which is comparable with an initial luminance (L_0) of about 10,000 cd/m^2. The result clearly indicates that the amount of residual water has a big influence on lifetimes. In the discussion of their paper, they commented that the number of water molecules striking the substrate exceeded that of the Alq$_3$ molecules by more than ten times in device A. They concluded that the device durability can be ascribed to the incorporation of water in the Alq$_3$ layer.

Yamamoto et al. of Universal Display Corporation (USA) etc. also investigated the effect of ultra-high vacuum condition on the lifetime of phosphorescent OLED devices [4]. The phosphorescent OLED device under ultra-high vacuum (UHV) condition of 6.5×10^{-7} Pa shows only 6% luminance reduction after 5 hours of operation at 1880 cd/m^2, while the same phosphorescent OLED device under high vacuum (HV) condition of 7.6×10^{-6} Pa shows 11% of luminance reduction in 5 hours of operation at 1774 cd/m^2. They reported that the amount of water incorporated within the EML of the devices was estimated to be approximately 9×10^9 molecules/cm^2 for the UHV device and approximately 3×10^{12} molecules/cm^2 for the HV device, which corresponds to 10^{-4} mol% and 0.05 mol%, respectively. It is obvious that the incorporated water plays an important role in degradation. They wrote that the typical electric field (~10^6 V/cm) applied to an OLED is large enough to induce electrochemical reduction of water, generating hydroxide and hydrogen ions, OH$^-$ and H$^+$ during device operation. They speculate that the addition of small quantities of ionic materials into the EML in the form of electrochemically reduced water induces degradation processes.

They also investigated the effect of water in the green phosphorescent OLED device. To investigate the primary device degradation impact, water was locally doped at precise positions in green PHOLEDs with the structure of ITO/HATCN/a-NPD/CBP:Ir(ppy)$_3$/BAlq/LiF/Al. From the six cases, they found that the α-NPD/EML interface is very sensitive to water. They propose that any chemically reactive ionic agents, OH$^-$ and H$^+$ in the α-NPD/EML interface recombination zone might react with the EML materials and act as quenchers, shortening device lifetimes.

6.2 Wet Processes

This section describes wet process for fabricating OLED devices, describing techniques such as ink-jet printing, nozzle printing, relief printing, and spray.

Figure 6.6 summarizes several wet processes, largely divided to two methods: processes for coating without fine patterning and processes for making fine patterns. The coating processes

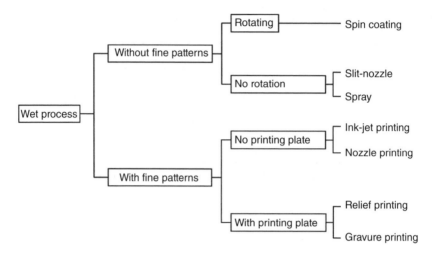

Figure 6.6   Classification of wet processes for OLED fabrications

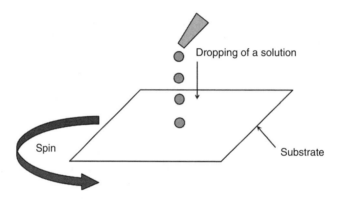

Figure 6.7   Schematic illustration of spin-coating

without fine patterning are used for organic film deposited on almost the whole device area. Therefore, these coating processes can be used in white OLEDs for OLED lighting and OLED displays with color filter and in non-emissive common organic layers such as HIL or HTL in OLED displays. On the other hand, organic layers for side-by-side RGB emitting layers require fine patterning. In such cases, fine patterning processes are required.

A typical technology for coating without fine patterning is spin-coating, which is well known and has often been used in the film preparation from solutions due to its advantage of uniform layer formation. As is shown in Fig. 6.7, in the spin-coating technique, solutions containing solutes are dropped onto a substrate, and then the substrate is rotated at high speeds, such as 1000 ~ 3000 rpm (revs per minute). While the spin-coating process is very convenient, especially for experimentation, it is not suitable for mass production because a lot of material is wasted in the spin-coating process.

Another typical coating process without fine patterns is slit-nozzle coating, which was originally developed for large size LCD production but can also be utilized in OLED wet processes.

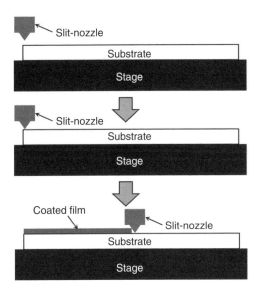

Figure 6.8   Schematic illustration of slit-coating

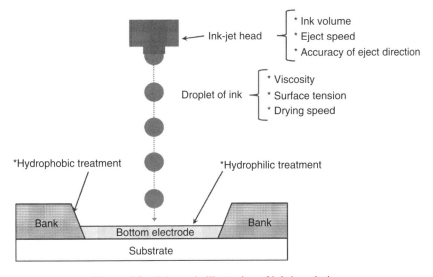

Figure 6.9   Schematic illustration of ink-jet printing

Figure 6.8 shows a schematic illustration of slit-coating. The slit-nozzle with a certain width ejects a solution at a controlled volume and speed.

Such coating processes are useful not only for experiments but also for production without fine patterning. Therefore, they can be applied to OLED lighting.

On the other hand, for obtaining an RGB side-by-side sub-pixel arrangement, fine patterning processes are required. One of the typical technologies is ink-jet process. Figure 6.9 shows the fundamental technique of the ink-jet process, along with some required properties.

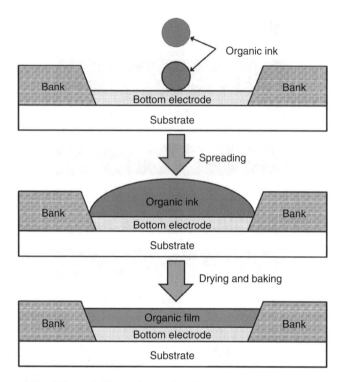

Figure 6.10   Schematic illustration of film formation process in ink-jet printing

The piezo-based multi-ink-jet nozzle ejects small droplets of ink to target places such as the sub-pixels of OLED displays. The volume of the droplets is several pico-liters or several tens of pico-liters, mainly determined by the ink-jet nozzle size. Since the distance between the ink-jet nozzle and the substrate is usually much larger than the size of the sub-pixel, the directional accuracy of ink ejection is very significant. In addition, accurate volume control and stable ejecting speed are also required for the ink-jet nozzle because they influence the fluctuation of film thickness and the process repeatability.

Figure 6.10 shows a schematic illustration of the film formation process in ink-jet printing. The dropped solution is elongated on the substrate because the viscosity of the solution should be low in order to give a good ejection. Therefore, a bank structure is usually required for keeping the solution within the sub-pixels. In addition, in order to obtain good spreading of the solution, the surface conditions of the substrate and the bank are very important. The surface treatment usually consists of two plasma steps, which are hydrophilic treatment of the surface and hydrophobic treatment of the bank. A delicate optimization is required to obtain wetting substrate surface and non-wetting bank, and to obtain a flat layer in the pixel.

The ink formulation also plays a significant role in jetting stability, jetting accuracy, and dried film flatness after the ink-jetting, etc. When the ink is dried, the film tends to give so-called coffee stains. One way for solving the coffee stain is using a mixture of solvents with low boiling point and high boiling point, respectively. By optimizing the solvent composition and surface conditions of the substrate and bank, uniform films can be obtained.

Table 6.2   Several prototype AM-OLED displays fabricated by ink-jet printing

	Size (diagonal)	Pixel number	Resolution	Driving	Year	Ref.
Seiko Epson	2.5″	200×150	100 ppi	LTPS-TFT	2001	[5]
Seiko Epson	2.1″		130 ppi	LTPS-TFT	2002	[6]
TMD	17″	1280×768	88 ppi	LTPS-TFT	2002	[7]
Philips	2.6″	220×176	107 ppi	LTPS-TFT	2004	[8]
Philips	13″	576×324	154 ppi	LTPS-TFT	2004	[9]
Casio	2.1″	160×128	101 ppi	a-Si-TFT	2004	[10]
Samsung	7.0″	480×320	82 ppi	a-Si-TFT	2005	[11]
Sharp	3.6″	640×360	202 ppi	CG-silicon TFT	2006	[12]
AU Optronics	65″	1920×1080	34 ppi	a-ITGO TFT	2014	[13]

TMD: Toshiba Matsushita Display Technology Co., Ltd

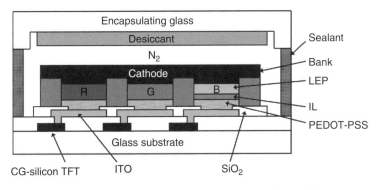

LEP: Light emitting polymer
IL: Interlayer

Figure 6.11   The device structure of a 3.6. full-color polymer OLED with 202 ppi [12]

Several prototype AM-OLED displays fabricated by ink-jet printing are summarized in Table 6.2.

The highest resolution of ink-jet printing was reported by Gohda et al. of Sharp Corporation in 2006 [12]. They developed a 3.6″ full-color polymer OLED with 202 ppi. The device structure is shown in Fig. 6.11. They utilized a CG-silicon (continuous grain silicon) [14] backplane with active matrix circuits, where the CG-silicon is classified as a modified low temperature poly-silicon (LTPS). They used two-layer structures with a PEDOT:PSS and an emitting polymer layer for red and green pixels. On the other hand, the blue pixel has a three-layer structure with an interlayer for obtaining a long lifetime. Bank structures are fabricated by using photosensitive organic film.

The pixel design is shown in Fig. 6.12. The pixel pitch is 42 μm for 202 ppi. The three organic layers of PEDOT:PSS, interlayer, and emitting polymer were fabricated by the ink-jet printing with a 7pl ink-jet head. Since the droplet size is about 23.7 μm, the required drop placement accuracy is about ±5.4 μm.

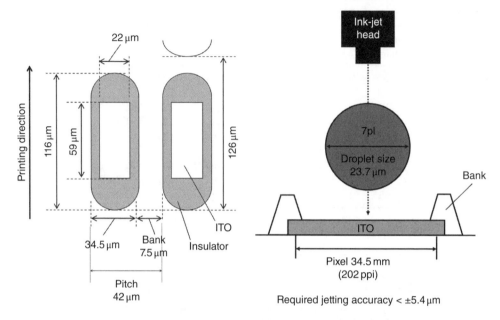

Figure 6.12 The pixel design of a 3.6. full-color polymer OLED with 202 ppi [12]

(a)   (b)

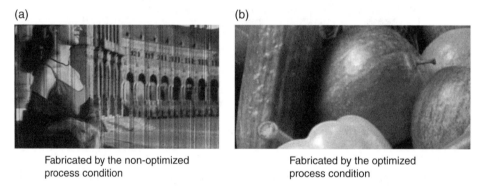

Fabricated by the non-optimized   Fabricated by the optimized
process condition   process condition

Figure 6.13 A picture of a 3.6. active-matrix full-color polymer OLED with 202 ppi [12]. (a) fabricated by non-optimized condition. (b) fabricated by optimized condition

As described above, to realize high resolution patterning such as 202 ppi, quite accurate patterning technology is required. If the process condition was not well optimized, display image quality is not good, as is shown in Fig. 6.13(a). A lot of defects appear due to the problems of ink-jet printing. Process conditions of hydrophilic treatment on ITO surface, hydrophobic treatment on bank surface, bank materials, ink-jet machine, ink formulation, etc. should be optimized. Figure 6.13(b) shows a display image obtained after the optimization of various parameters.

The largest OLED panel fabricated by ink-jet printing is a 65″ AM-OLED display reported by Chen et al. of AU Optronics Corporation (Taiwan) [13].

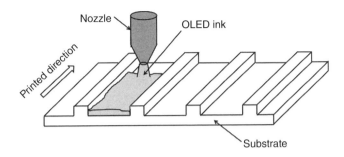

Figure 6.14 Schematic illustration of nozzle printing [15, 16]

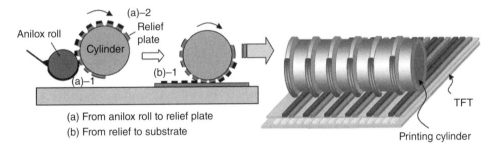

Figure 6.15 Schematic illustration of relief printing [18]

Other wet processes have also been reported for OLED fabrications. For example, DuPont Displays reported the continuous nozzle printing technology, which has been developed together with Dainippon Screen Manufacturing Co., Ltd (Japan) [15, 16]. The schematic illustration of the nozzle printing is shown in Fig. 6.14. Continuous nozzle printing utilized a laminar liquid jet that issues from a fixed orifice and then impinges on the substrate. The printed process operates by continuous moving the liquid jet across the substrate in alignment with previously defined wetting and non-wetting area. They presented a 4.4″ AM-OLED display fabricated by the nozzle printing [16].

Takeshita and Onohara et al. of Toppan Printing Co., Ltd (Japan) reported on the utilization of a relief printing [17, 18]. This method is a kind of direct printing with a relief plate, as shown in Fig. 6.15. The required patterns are formed as convex patterns on the relief plate. In the first step, the inks supplied from the ink pan to the anilox roll are transferred to the convex parts of the relief plate with the designed pattern. In the second step, the inks on the convex parts of the relief plate are transferred to the substrate. The printed film thickness is controlled by the amount of inks maintained on the anilox roll. They also reported that the uniformity and form of the printed film are influenced by the ink properties at the time of the transfer to the substrate. They fabricated a 5″ prototype of full-color polymer OLED.

High throughput roll-to-roll printing technologies are attractive for the possibility of low-cost OLED production. Various printing techniques have been investigated. Kopola et al. of VTT (Finland) applied gravure printing technology to OLED [19]. Hast et al. of VTT Technical Research Centre of Finland fabricated OLED devices by using a roll-to-roll gravure printing technology [20].

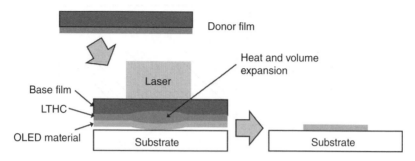

Figure 6.16   The schematic illustration of the LITI (laser induced thermal imaging [25]

Takakuwa et al. of Japan Chemical Innovation Institute (JCII) etc. investigated micropatterning by microcontact printing using a poly(dimethylsiloxane) (PDMS) elastomer stamp, which is fabricated using the nanoimprint method [21]. The patterned pixel size is $400 \times 2000\,\mu m$.

Hsieh et al. of National Taiwan University et al. (Taiwan) proposed and developed an air-bubble coating method for OLED devices as a novel pattern-coating technology [22].

Panasonic reported a 55″ 4K2K OLED display fabricated by an all-printing method [23]. They utilized an IGZO backplane and a top emitting OLED device structure. The specifications are pixel number of 3850×2160 dots, peak luminance of $500\,cd/m^2$, contrast ratio of 1,000,000:1, color gamut of NTSC 110%, gray scale of 10 bits, etc.

6.3   Laser Processes

The patterning processes using a laser have been proposed and developed. This section describes two laser patterning processes: the laser induced thermal imaging technology proposed by Samsung and 3 M, and the laser-induced pattern-wise sublimation (LIPS) technology proposed by Sony.

The laser induced thermal imaging (LITI) was first proposed and reported by Lee et al. of Samsung SDI Co. Ltd (South Korea) and 3 M Display Materials Technology Center (USA) [24]. The schematic illustration of LITI is shown in Fig. 6.16 [25].

The LITI process utilizes a donor film, a laser exposure system and a substrate. The donor film is a transparent film with a light-to-heat conversion (LTHC) layer. The LTHC layer absorbs irradiated light and converts it to heat. Lee et al. reported that dyes absorbing in the IR region of the spectrum and pigment materials such as carbon black and graphite can be used as the LTHC layer.

On the donor film, an organic material layer is deposited, and then the donor film is placed in intimate contact with a substrate and exposed by the laser system. In the area irradiated by the laser, a volume expansion of the LTHC layer occurs, and then the organic material is transferred to the substrate from the donor film.

The merits of using lasers are high resolution and flexibility in the format and size of the final images. Lee et al. reported that the overall position accuracy is typically less than $\pm 2.5\,\mu m$ [24].

Yoo et al. of Samsung SDI Co., Ltd developed a 2.6″ full-color VGA AM-OLED display with 302 ppi by using the LITI process with the top emitting structure [25]. They used an

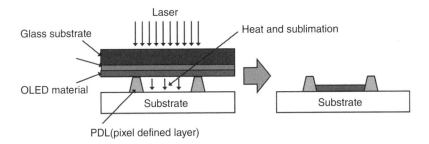

Figure 6.17 The schematic illustration of the LIPS (laser-induced pattern-wise sublimation) [26]

Table 6.3 27.3″ full-color AM-OLED display by using the LIPS [26]

Display size	27.3″ diagonal
Number of pixels	1920×1080 (full HD)
Pixel pitch	315×315 μm
Brightness	All white: 200 cd/m²
	Peak: >600 cd/m²
Contrast ratio	>1,000,000:1

LTPS backplane consisting of a voltage compensating pixel circuit with six TFTs and one capacitor in each pixel.

On the other hand, the laser-induced pattern-wise sublimation (LIPS) technology was proposed and reported by Hirano et al. of Sony Corporation (Japan) [26]. The schematic illustration of LIPS is shown in Fig. 6.17.

Organic materials are deposited on the glass donor, covered with molybdenum absorption layer. On the other hand, the OLED substrate should have a pixel defined layer (PDL).

The OLED substrate and the glass donor are set with a certain gap in a vacuum chamber, not being exposed to the air. After mechanical alignment of the substrate to the laser head, the laser beam scans and heats the designed position of the glass donor. They used an 800 nm diode laser as the radiation source. The width of the laser beam is adjusted in accordance with the width of the transferred pattern. OLED materials are precisely transferred to a substrate from glass donors by a scanning laser beam. Glass donors can be reused. They reported that the pattern width variation is within ±2 μm.

They fabricated a 27.3″ full-color AM-OLED display by using the LIPS technology. The specification of the 27.3″ display is shown in Table 6.3.

References

[1] H. Kubota, S. Miyaguchi, S. Ishizuka, T. Wakimoto, J. Funaki, Y. Fukuda, T. Watanabe, H. Ochi, T. Sakamoto, T. Miyake, M. Tsuchida, I. Ohshita, T. Tohma, *Journal of Luminescence*, **87**, 56–60 (2000).

[2] E. Fujimoto, H. Daiku, K. Kamikawa, E. Fujimoto, Y. Matsumoto, *SID 10 Digest*, 46.3 (p. 695) (2010).

[3] T. Ikeda, H. Murata, Y. Kinoshita, J. Shike, Y. Ikeda, M. Kitano, *Chem. Phys. Lett.*, **426**, 111–114 (2006).

[4] H. Yamamoto, M. S. Weaver, H. Murata, C. Adachi, J. J. Brown, *SID 2014 Digest*, 52.3 (p. 758) (2014).

[5] S. K. Heeks J. H. Burroughes, C. Town, S. Cina, N. Baynes, N. Athanassopoulou, J. C. Carter, *SID 01 Digest*, 31.2 (2001).

[6] T. Funamoto, Y. Matsueda, O. Yokoyama, A. Tsuda, H. Takeshita, S. Miyashita, *SID 02 Digest*, 27.5 L (p. 899) (2002).

[7] M. Kobayashi, J. Hanari, M. Shibusawa, K. Sunohara, N. Ibaraki, *Proc. IDW'02*, AMD3-1 (p. 231) (2002).

[8] M. Fleuster, M. Klein, P. v. Roosmalen, A. d. Wit, H. Schwab, *SID 04 Digest*, 44.2 (p. 1276) (2004).

[9] N. C. van der Vaart, H. Lifka, F. P. M. Budzelaar, J. E. J. M. Rubingh, J. J. L. Hoppenbrouwers, J. F. Dijksman, R. G. F. A. Verbeek, R. van Woudenberg, F. J. Vossen, M. G. H. Hiddink, J. J. W. M. Rosink, T. N. M. Bernards, A. Giraldo, N. D. Young, D. A. Fish, M. J. Childs, W. A. Steer, D. Lee, D. S. George, *SID 04 Digest*, 44.4 (2004).

[10] T. Shirasaki, T. Ozaki, K. Sato, M. Kumagai, M. Takei, T. Toyama, S. Shimoda, T. Tano, *SID 04 Digest*, 57.4 L (p. 1516) (2004).

[11] D. Lee, J.-K. Chung, J.-S. Rhee, J.-P. Wang, S.-M. Hong, B.-R.Choi, S.-W.Cha, N.-D.Kim, K. Chung, H. Gregory, P. Lyon, C. Creighton, J. Carter, M. Hatcher, O. Bassett, M. Richardson, P. Jerram, *SID 05 Digest*, P-66 (p. 527) (2005).

[12] T. Gohda, Y. Kobayashi, K. Okano, S. Inoue, K. Okamoto, S. Hashimoto, E. Yamamoto, H. Morita, S. Mitsui, M. Koden, *SID 06 Digest*, 58.3 (2006); M. Koden, Y. Hatanaka, Y. Fujita, Y. Kobayashi, E. Yamamoto, K. Ishida, S. Mitsui, *3rd Japanese OLED Forum*, S7-1 (2006).

[13] P.-Y. Chen, C.-L. Chen, C.-C. Chen, L. Tsai, H.-C. Ting, L.-F. Lin, C.-C. Chen, C.-Y. Chen, L.-H. Chang, T.-H. Shih, Y.-H. Chen, J.-C. Huang, M.-Y. Lai, C.-M. Hsu, Y. Lin, *SID 2014 Digest*, 30.1 (p. 396) (2014).

[14] H. Sakamoto, N. Makita, M. Hijikigawa, M. Osame, Y. Tanada, S. Yamazaki, *SID 00 Digest*, 53.1 (p. 1190) (2000).

[15] W. F. Feehery, *SID 07 Digest*, 69.1 (p. 1834) (2007).

[16] R. Chesterfield, A. Johnson, C. Lang, M. Stainer, J. Ziebarth, *Information Display*, 1/11, p. 24 (2011).

[17] K. Takeshita, H. Kawakami, T. Shimizu, E. Kitazume, K. Oota, T. Taguchi, I. Takashima, *Proc. IDW/D'05*, OLED2-2 (p. 597) (2005).

[18] J. Onohara, K. Mizuno, Y. Kubo and E. Kitazume, *SID 2011 Digest*, 62.2 (2011).

[19] P. Kopola, M. Tuomikoski, R. Suhonen, A. Maaninen, *Thin Solid Films*, **517**, 5757–5762 (2009).

[20] J. Hast, M. Tuomikoski, R. Suhonen, K.-L. Väisänen, M. Välimäki, T. Maaninen, P. Apilo, A. Alastalo, A. Maaninen, *SID 2013 Digest*, 18.1 (p. 192) (2013).

[21] A. Takakuwa, M. Misaki, Y. Yoshida, K. Yase, *Thin Solid Films*, **518**, 555–558 (2009).

[22] Y.-W. Hsieh, P.-T. Pan, A.-B. Wang, L. Tsai, C.-L. Chen, P.-Y. Chen, W.-J. Cheng, *SID 2014 Digest*, 30.4 (p. 407) (2014).

[23] H. Hayashi, Y. Nakazaki, T. Izumi, A. Sasaki, T. Nakamura, E. Takeda, T. Saito, M. Goto, H. Takezawa, *SID 2014 Digest*, 58.3 (p. 853) (2014).

[24] S. T. Lee, J. Y. Lee, M. H. Kim, M. C. Suh, T. M. Kang, Y. J. Choi, J. Y. Park, J. H. Kwon, H. K. Chung, J. Baetzold, E. Bellmann, V. Savvate'ev, M. Wolk, S. Webster, *SID 02 Digest*, 21.3 (p. 784) (2002).

[25] K.-J. Yoo, S.-H. Lee, A.-S. Lee, C.-Y. Im, T.-M. Kang, W.-J. Lee, S.-T. Lee, H.-D. Kim, H.-K. Chung, *SID 05 Digest*, 38.2 (p. 1344) (2005); S. T. Lee, M. C. Suh, T. M. Kang, Y. G. Kwon, J. H. Lee, H. D. Kim and H. K. Chung, *SID 07 Digest*, 53.1(p. 1158) (2007).

[26] H. Hirano, K. Matsuo, K. Kohinata, K. Hanawa, T. Matsumi, E. Matsuda, R. Matsuura, T. Ishibashi, A. Yoshida and T. Sasaoka, *SID 2007 Digest*, 53.2 (p. 1592) (2007).

7

Performance of OLEDs

Summary

Performance of OLED devices is very important not only in scientific studies but also in applications to practical devices and commercial products. While various characteristics and parameters of OLED devices have already been described in the previous chapters, this chapter explains typical characteristics and parameters of OLED devices and describes current performances of OLEDs. These characteristics are I–V–L characteristics, efficiencies, and lifetimes. In addition, this chapter describes the temperature elevation phenomenon in OLED devices, which is also important in practical applications because such temperature elevation is closely related to lifetime.

Key words

performance, I–V–L characteristics, efficiency, lifetime, temperature elevation

7.1 Characteristics of OLEDs

In the fundamental scientific field, one of the important characteristics of OLED devices is quantum efficiency. There are two quantum efficiencies in OLEDs: internal quantum efficiency η_{int} (IQE) and external quantum efficiency η_{ext} (EQE).

The internal quantum efficiency η_{int} is defined as the ratio of the number of emitted photons to the number of recombined holes and electrons.

As described in Section 3.4, the number of emitted photons is not the same as the number of observed photons outside the OLED devices. The out-coupling efficiency η_{out} is defined as the ratio of the number of observed photons outside the OLED device and the number of emitted photons inside the OLED devices. Therefore, the η_{ext} is defined as the product of the

OLED Displays and Lighting, First Edition. Mitsuhiro Koden.
© 2017 John Wiley & Sons, Ltd. Published 2017 by John Wiley & Sons, Ltd.

internal quantum efficiency η_{int} and the out-coupling efficiency η_{out} as is described by the following equation:

$$\eta_{ext}\left(EQE\right)=\eta_{out}\times\eta_{int}\left(IQE\right)$$

In applied science and practical technology fields, the I–V–L (current–voltage–luminance) characteristic is more realistically significant. However, it should be noted that the L (luminance) includes the luminosity function, which is related to human eyes. In addition, it should be noticed that the luminance depends on the observation direction with flat and curved OLED devices, while the luminance is often measured only in the normal direction of flat OLED devices. Since the emission characteristics are dependent upon the emission profile of OLED devices, it should be noted that emission characteristics cannot be described only by luminance.

Nevertheless, the I–V–L characteristic is convenient and often used. Examples of I–V–L characteristics and emission spectra are shown in Figs 7.1 to 7.3.

Figure 7.1 shows a typical I–V characteristic of an OLED device. As shown in Figs. 7.1 (a) and (b), two types of representations are often used. In (a), the current on the vertical axis is logarithmically represented, while in (b) the current on the vertical axis is represented by a linear scale. The representation using a logarithmic scale such as (a) has two advantages. First, this clearly indicates the turn-on voltage. Indeed, Fig. 7.1 (a) clearly indicates that the turn-on voltage is 2.0 V, though the turn-on voltage is not so clear in Fig. 7.1 (b). Second, the representation using logarithmical scale such as (a) clearly shows the leakage level of the OLED device. Normal OLED devices tend to show a lower current level than 10^{-4} mA/cm² under the turn-on voltage if they are well fabricated. However, if OLED devices have some defects and/or other causes of leakage, the I–V characteristics show a higher current level than 10^{-4} mA/cm². Therefore, the representation using logarithmic scale can give information about the device failure. However, it is difficult to obtain such information about device failure from the representation using a linear scale such as (b).

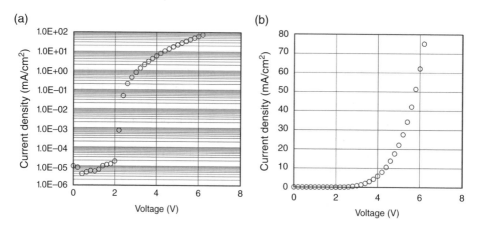

Figure 7.1 I–V characteristics of an OLED device with the structure of glass/ITO(150 nm)/ MoO₃(10 nm)/α-NPD(40 nm)/Alq₃(30 nm)/DPB:Liq(25 wt%)(43.5 nm)/Al(100 nm)

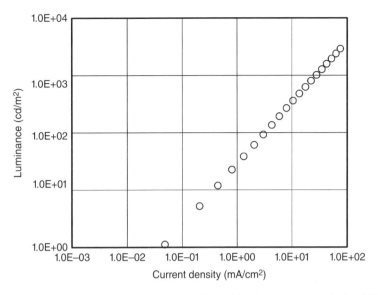

Figure 7.2 L–I characteristics of an OLED device with the structure of glass/ITO(150 nm)/ MoO$_3$(10 nm)/α-NPD(40 nm)/Alq$_3$(30 nm)/DPB:Liq(25 wt%)(43.5 nm)/Al(100 nm). (Same device as Fig. 7.1.)

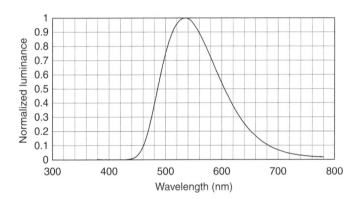

Figure 7.3 Emission spectrum of OLED devices with the structure of glass/ITO(150 nm)/ MoO$_3$(10 nm)/α-NPD(40 nm)/Alq$_3$(30 nm)/DPB:Liq(25 wt%)(43.5 nm)/Al(100 nm). (Same device as Fig. 7.1.)

Figure 7.2 is a typical L–I characteristic for the same device as Fig. 7.1. Both coordinates are commonly given a logarithmic scale. In the L–I characteristics, the graphs tend to show an almost linear relationship.

The emission spectrum is also important data because color is an important parameter in practical applications. In addition, emission spectra can give information about the inside of OLED devices. They provide information about layer thickness, recombination site, emitting species, excitons, etc., which influence emitting spectra. A typical example of an emission spectrum is shown in Fig. 7.3.

Emission efficiency is very important because it is related to power consumption of OLED devices. Two emission efficiencies are often used. They are current efficiency (cd/A) and

Table 7.1 Typical performances of R, G, and B OLED devices

Type of OLED		Color	Efficiency	CIE	LT_{50}	Affiliation	Ref.
			cd/A				
Evaporation type	Small molecule	B	4.7	(0.148, 0.062)	>10,000 @500 cd/m²	Merck	[1]
				Light blue	100 Khrs @300 cd/m²	Konica Minolta	[2]
			50	(0.18, 0.42)	20,000 hrs	UDC	[3]
		G	85	(0.31, 0.63)	400,000 hrs	UDC	[3]
			93.3			LG Chem	[4]
		R	30	(0.64, 0.36)	900,000 hrs	UDC	[3]
			39.7	(0.666,0.334)		SEL	[5]
Solution type	Small molecule	B	6.2	(0.14, 0.14)	24,000	DuPont	[6]
		G	68.3	(0.32, 0.63)	125,000		[6]
		R	20.3	(0.65, 0.35)	200,000		[6]
	Polymer	B	12	(0.14, 0.12)	>10 k	Sumitomo	[7]
		G	76	(0.32, 0.63)	450 k	Chemical	[7]
		R	31	(0.62, 0.38)	350 k		[7]

power efficiency (lm/W). These are calculated from the I–V–L characteristics as shown in Figs 7.1 and 7.2. In OLED lighting, it is almost true that the power efficiency is directly related to power consumption. However, in OLED displays, the contribution of circuits and TFT drives to power consumption cannot be neglected. Therefore, the power efficiency is not directly related to the power consumption of OLED displays.

Several typical performance of red, green, and blue OLED devices are summarized in Table 7.1, while typical performance of white OLEDs will be described in Section 9.5.

While deep blue phosphorescent OLEDs that have a long lifetime have not yet been developed, light blue phosphorescent OLEDs have been reported as achieving long lifetime. Ito et al. of Konica Minolta developed a light blue phosphorescent OLED with EQE of 23% and half lifetime LT_{50} of 100,000 hours at 300 cd/m² [2].

7.2 Lifetime

Lifetime is one of the important characteristics of OLEDs. There are two types of lifetime: storage lifetime and driving lifetime. These two types of lifetime are summarized in Fig. 7.4.

Storage lifetimes are related to the degradation occurring under certain storage conditions without operation. In practical applications, the storage lifetime is defined as the period for which the products maintain acceptable performance. While storage lifetimes are influenced by various factors, storage lifetimes in actual cases are closely related to a protection from attacks by water vapor and oxygen. By the penetration of moisture and oxygen, so-called dark spots and dark areas appear, as shown in Fig. 7.4.

On the other hand, driving lifetimes are related to the phenomenon that luminance decreases under driving conditions.

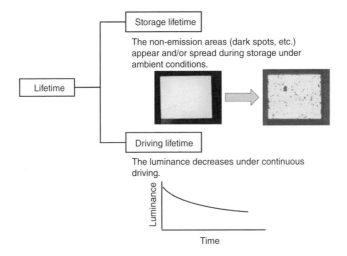

Figure 7.4 Two types of lifetimes in OLEDs

Table 7.2 Acceleration factors of several measurement conditions [8]

Condition	25 °C/60%RH	40 °C/90%RH	60 °C/90%RH	85 °C/90%RH
Acceleration factor	1	7	39	260

7.2.1 Storage Lifetime

Generally, storage lifetimes in OLEDs are related to such phenomena that non-emission areas (dark spots, dark areas, etc.) appear and spread during storage under ambient conditions. This degradation is caused by the attack of moisture and oxygen penetrating into OLED devices from the ambient air. Therefore, storage lifetimes are closely related to encapsulating technologies, as is described in Section 5.7.

In commercial products, the acceptable level of appearance of non-emission area is decided, based on general concepts of commercialized products.

Storage lifetime is usually evaluated using accelerated measurement under high temperature and humidity conditions. The often used accelerated conditions are 60 °C/90%RH, 85 °C/90%RH, etc. Arai et al. reported the acceleration factors of several measurement conditions as shown in Table 7.2 [8].

7.2.2 Driving Lifetime

Driving lifetime is related to luminance reduction under the operation of OLED devices. Lifetime curves and several definitions of lifetime are schematically illustrated in Fig. 7.5. In the driving lifetime measurements, luminance reduction is usually measured under constant current drive. In addition, since the OLED driving also gives rise to an increase in driving voltage, the increase in driving voltage is also often evaluated.

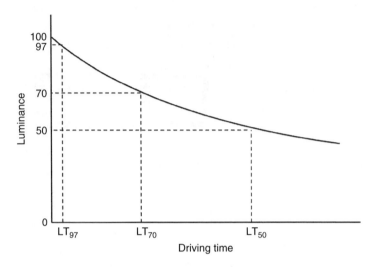

Figure 7.5   Schematic illustration of a lifetime curve and several definitions of lifetimes

Classically, the driving lifetime has often been defined by half lifetime (LT_{50}). The half lifetime is defined by the time at which the luminance reaches half of the initial luminance (L_0) under a certain driving condition.

However, LT_{50} is often inadequate in practical lighing applications because the half luminance is too dark compared with the initial luminance. Therefore LT_{70} is often used in OLED lightings.

In display applications, the situation is more complicated because a burn-in phenomenon occurs. Figure 7.6 shows a schematic illustration of the burn-in effect. In this case, it is assumed that a fixed pattern of "OLED" in a black background is displayed for a certain period, and then a picture image is displayed. Since the accumulated driving period of pixel B, on which the fixed pattern of "OLED" emits, is longer than that of pixel A, the luminance reduction in pixel B is larger than that in pixel A. Therefore, when a picture image is displayed, the fixed pattern of "OLED" is burned-in.

This phenomenon often occurs because mobile phone and digital TV often display fixed patterns. Therefore, LT_{97} etc. are often used as the lifetime for display applications.

Lifetime is often evaluated by such acceleration tests as high temperature acceleration, high luminance acceleration, etc. High luminance acceleration is the most usually used method. A typical example of high luminance accelerated lifetime estimation is shown in Fig. 7.7. The relationship between lifetime and initial luminance is often represented by the following equation, where T is lifetime and L_0 is initial luminance.

$$T \propto \left(1/L_0\right)^n$$

The n is considered to be 1 if we suppose that the luminance reduction is in proportion to accumulated driving current. However, the actual value of n is not usually 1. Empirically, it is known that the n of small molecular OLEDs is $1.3 \sim 1.5$ and the n of polymer OLEDs is $1.8 \sim 2.0$.

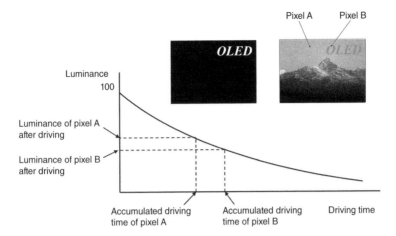

Figure 7.6   Schematic illustration of burn-in effect

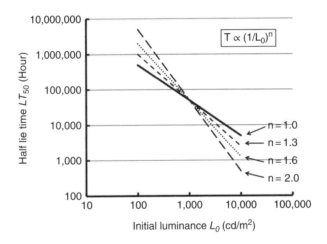

Figure 7.7   A typical example of high luminance accelerated lifetime estimation

There are several reports on lifetime evaluation under high temperature acceleration. Parker et al. reported that the half lifetime at 25 °C is more than 100 times longer than at 85 °C [9]. On the other hand, Koden and coworkers at Sharp Corporation reported the temperature dependence of lifetimes in polymer OLEDs, indicating that the lifetimes at 20 °C are only about several or tens of times longer than those at 80 °C [10]. The data reported by Ishii et al. shows similar acceleration factors [11, 12]. The discrepancy of acceleration factors may be explained by supposing that the efficiency of OLEDs affects the acceleration factor because the current heating phenomenon gives rise to temperature elevation of OLED devices.

7.3 Temperature Measurement of OLED Devices

It is well known that driving of OLED devices induces temperature elevation, giving rise to a reduction of lifetime. However, the temperature measurement of OLED devices is not so easy because the temperature of emitting organic material is not the same as the temperature of the substrates of OLED devices.

Various evaluation technologies have been investigated and reported. There are many reports utilizing spectral change of organic materials induced by temperature elevation under driving. They include analysis of emission spectra [13, 14], IR imaging [15], scanning thermal microscopy (SThM) [16], and analysis of Raman spectra [17, 18].

However, temperature estimation methods using such spectral analyses have several restrictions. First, analysis equipment and analysis technique are required. Second, the types of OLED devices are restricted because the OLEDs need to have adequate materials showing adequate spectra for such estimation.

By contrast, one useful and easy way is the temperature evaluation method using the temperature dependence of the I–V characteristics of OLED devices [10]. As the first step in this method, the temperature dependence of the I–V characteristics are measured, as shown in Fig. 7.8. The current density at every applied voltage increases with increasing temperature. As the second step, the change of the current density is measured under driving. As an example, Fig. 7.9(a) shows the change of the current density under continuous driving with constant applied voltage in the OLED device used in Fig. 7.8. In continuous operation, the current density increases, owing to the temperature elevation. By comparing the data of Fig. 7.8 with the data of Fig. 7.9(a), the temperature elevation is estimated as shown in Fig. 7.9(b).

Since I–V characteristics of OLED devices are easily evaluated, this method can be applied to the evaluation of the temperature elevation in all OLED devices. Figure 7.10 indicates two

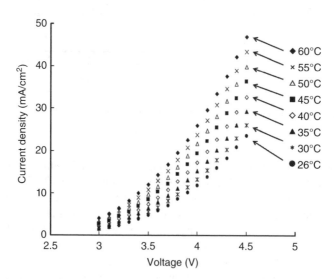

Figure 7.8 The temperature dependence of the I–V characteristics of an OLED device with the structure of ITO/PEDOT-PSS/LEP/Ca/Ag

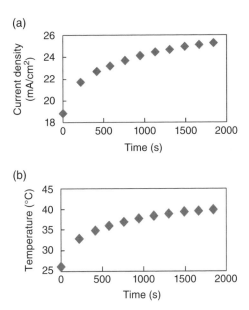

Figure 7.9 The current change and the temperature elevation of the OLED device under continuous operation with constant applied voltage of 4.3 V [10]. The device structure is ITO/PEDOT-PSS/LEP/Ca/Ag (same device as Fig. 7.8). The initial luminance was 900 cd/m²

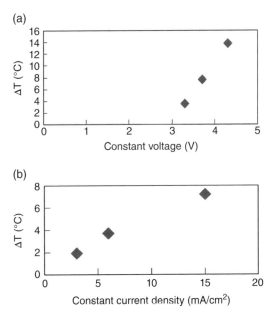

Figure 7.10 The elevated temperature ΔT of LEP in the OLED after continuous operation for 30 minutes [10]. The device structure is ITO/PEDOT-PSS/LEP/Ca/Ag. (same device with Fig. 7.8). (a) Constant voltage operation. (b) Constant current operation

significant pieces of information. First, the temperature elevation under constant current drive is smaller than that under constant voltage drive, in spite of the initial luminance being the same. This is attributed to the fact that the temperature elevation induces an increase in supplied energy because the current increases due to the temperature elevation. On the other hand, in constant current drive, the applied energy decreases under driving because the current is constant and the voltage decreases due to the temperature elevation. The second piece of information obtained from Fig. 7.10 is that the temperature elevation is not so large. Specifically, in constant current operation, the elevated temperature is only 7 °C when the initial luminance is 900 cd/m^2 and the efficiency of the device is not so high (about 6 cd/A).

References

[1] H. Heil, L. Rodriguez, B. Burkhart, S. Meyer, S. Riedmueller, A. Darsy, C. Pflumm, H. Buchholz, E. Boehm, *SID 2014 Digest*, 35.1 (p. 495) (2014).
[2] H. Ito, K. Kiyama, H. Kita, *Proc. IDW'13*, OLED1–3 (p. 860) (2013).
[3] Home page of Universal Display Corporation: www.udcoled.com/default.asp?contentID=604 (March 2015).
[4] J. K. Noh, M. S. Kang, J. S. Kim, J. H. Lee, Y. H. Ham, J. B. Kim, M. K. Joo, S. Son, *Proc. IDW'08*, OLED3-1 (p. 161) (2008).
[5] H. Inoue, T. Yamaguchi, S. Seo, H. Seo, K. Suzuki, T. Kawata, N. Ohsawa, *SID Digest 2014*, 35.4 (p. 505) (2014).
[6] N. Herron, W. Gao, *SID 10 Digest*, 32.3 (p. 469) (2010).
[7] T. Yamada, Y. Tsubata, D. Fukushima, K. Ohuchi, N. Akino, *Proc. IDW'14*, OLED5-1 (2014).
[8] T. Arai, K. Konno, T. Miyasako, M. Takahashi, M. Nishikawa, K. Azuma, S. Ueno, M. Yomogida, *Proc. of 15th Japanese OLED Forum*, S7-4 (p. 49) (2012).
[9] I. D. Parker, Y. Cao, C. Y. Yang, *J. Appl. Phys.*, **85(4)**, 2441–2447 (1999).
[10] M. Koden, S. Okazaki, Y. Fujita and S. Mitsui, *Proc. IDW'03*, OEL3-3 (p. 1313) (2003).
[11] M. Ishii, Y. Taga, *Appl. Phys. Lett.*, **80(18)**, 3430–3432 (2002).
[12] T. Sato and M. Ishii, *Proc. 5th Japanese OLED Forum*, S7-1 (p. 37) (2007).
[13] N. Tessler, T. Harrison, D. S. Thomas and R. H. Friend, *Appl. Phys. Lett.*, **73(6)**, 732–734 (1998).
[14] J. M. Lupton, *Appl. Phys. Lett.*, **81(13)**, 2478–2480 (2002).
[15] X. Zhou, J. He, L. S. Liao, M. Lu, X. M. Ding, X. Y. Hou, X. M. Zhang, X. Q. He, and S. T. Lee, *Adv. Mater.*, **12**, 265 (2000).
[16] F. A. Boroumand, M. Voigt, A. D. G. Lidzey, Hammiche, G. Hill, *Appl. Phys. Lett.*, **84(24)**, 4890–4892 (2004); F. A. Boroumand, A. Hammiche, G. Hill, D. G. Lidzey, *Adv. Mater.*, **16**, 252 (2004).
[17] R. Iwasaki, M. Hirose and Y. Furukawa, *Jpn. J. Appl. Phys.*, **52**, 05DC16 (2013).
[18] T. Sugiyama, H. Tsuji, Y. Furukawa, *Chem. Phys. Lett.*, **453**, 238–241 (2008).

8

OLED Display

Summary

Display is one of the important application fields for OLEDs because OLEDs have several excellent features as displays. Before describing OLED display technologies, this chapter first describes the unique features of OLED displays in comparison to liquid crystal displays (LCDs), which are used for the majority of current displays.

OLED displays are divided to static drive, passive-matrix drive and active-matrix drive by the driving methods. Static drive can allow only very low information contents. Passive-matrix OLEDs (PM-OLED) can allow medium information contents, having been applied, for example, to car audio, sub-display of mobile phone, small size main display of mobile phone. On the other hand, active-matrix OLEDs (AM-OLED), which are usually driven by thin film transistors (TFTs), can allow large information contents. Therefore, AM-OLEDs can be applied to wide ranges of applications such as mobile phone, smart phone, tablet display, and TV. In AM-OLEDs, the TFT technologies are different from those for AM-LCDs. This chapter describes the TFT materials and TFT circuit technologies for AM-OLED displays.

Key words

display, static drive, passive-matrix drive, active-matrix drive, thin film transistor, circuit, full color

OLED Displays and Lighting, First Edition. Mitsuhiro Koden.
© 2017 John Wiley & Sons, Ltd. Published 2017 by John Wiley & Sons, Ltd.

8.1   Features of OLED Displays

Display is one of the important application fields of OLEDs because they can realize excellent picture quality, which means high (ultimate) contrast ratio, high brightness, peak luminance characteristic, wide viewing angle, fast response time, high color reproducibility, high resolution, and high information contents. In addition, OLED displays have such attractive features as being thin and light weight, and have high potential for low power consumption and low production cost.

On the other hand, the current major display is LCD, so it is important to compare OLED displays with LCDs.

LCD usually utilizes an optical shutter effect induced by a combination of a liquid crystal device and polarized light. Since the shutter effect of LCDs tends to be disturbed by fluctuating molecular orientation and imperfect molecular switching, etc., the contrast ratio of LCD is restricted. In addition, since LCD usually utilizes the switching of anisotropic molecular orientation of liquid crystal molecules, the viewing angle characteristic also has problems such as contrast ratio reduction and color change at the inclined direction. But since OLED is a self-emissive device, the contrast ratio is fundamentally constant and is very high (as higher as 1,000,000:1) in every direction in actual devices, and the color change at an inclined angle is very small.

Also, the response time of LCDs is not as good, being around several milliseconds, since common LCDs utilize the orientational switching of liquid crystal molecules with the anisotropic molecular shape. OLEDs, however, utilize only charge injection, excitation, and emission within very thin organic films of around 100 nm. No dynamic movement of molecules occurs, so the switching speed of OLEDs is very high, being of the order of microseconds. The sharpness of moving pictures in displays is not directly related to the response time of the devices but is also dependent upon the drive scheme [1], but the advantage of OLEDs still remains.

Since the maximum brightness of LCDs is restricted by the luminance of the backlight, the highest brightness is not dependent upon picture images. On the other hand, since the OLED is a current driven and self-emissive display, it can realize peak luminance characteristics such that the brightness of a peak white point with black background is higher than that of the whole white image.

In addition, by optimizing organic materials and layer structure, etc., OLEDs can realize high brightness and good color reproducibility.

Since OLEDs do not require a backlight system, unlike LCDs, the OLED can be thinner and lighter than the LCD.

The power consumption and production costs of OLEDs are still issues at present, but OLEDs have the possibility of lower power consumption and lower production cost than LCD, because OLEDs do not need a color filter, backlight, or a pair of polarizers.

Finally, it should be noted that OLEDs can be flexible. Therefore, the flexible OLED display is being actively developed.

8.2   Types of OLED Displays

OLED displays are classified by the driving methods, as shown Fig. 8.1. Driving methods of OLED displays are divided into segment drive and dynamic drive.

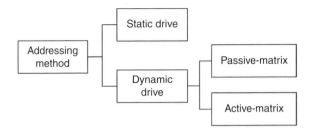

Figure 8.1 Classification of OLED displays

In the segment drive scheme, each emission area or pixel is driven individually. Therefore, the driving technology is not so complicated. However, since the number of emission areas or pixels is consistent with the number of segments of electrodes, large information content is practically impossible.

On the other hand, dynamic drive schemes can realize large information content because they can control the emission of pixels by supplying signals to lines of x-direction and y-direction. Dynamic drive OLEDs can be divided to passive-matrix OLED (PM-OLED) and active-matrix OLED (AM-OLED), as illustrated in Fig. 8.2.

Passive-matrix OLEDs are also called, for example, simple-matrix OLED, multiplex-drive OLED. In passive-matrix OLEDs, the row electrode (x-line) is time-sequentially addressed single line by single line. The emission intensity of each pixel is controlled by a signal from the corresponding column electrode (y-line).

On the other hand, active-matrix OLEDs control the emission of each pixel usually by using a thin film transistor (TFT) which is attached to each pixel located at the point of intersection of the x- and y-lines. The row electrode (x-line) is time-sequentially addressed line by line. The emission intensity of each pixel is controlled by a signal from the corresponding column electrode (y-line). While the number of TFTs at each pixel is usually only one in active-matrix LCDs, there have to be a number of TFTs in active-matrix OLEDs, because LCD is a capacitance device whereas OLED is a current device. Several TFT materials such as amorphous-Si (a-Si), poly-Si, single crystal Si, metal oxide, and organic semiconductor have been investigated for active-matrix OLEDs, and some of them have been made commercially available. The detail will be described later.

A comparison of passive-matrix OLED and active-matrix OLED is summarized in Table 8.1. The advantage of PM-OLED is the simple device structure because PM-OLEDs require no fabrication process for TFTs. However, PM-OLEDs have several restrictions, as shown in Table 8.1, being attributed to the short emission period of each pixel. Since the row electrode is time-sequentially selected, the emission period of pixels on the selected row electrode is restricted. For example, if we suppose 100 row lines and a panel luminance of $500 \, \mathrm{cd/m^2}$, the required luminance of each pixel is $50,000 \, \mathrm{cd/m^2}$, that is 100 times $500 \, \mathrm{cd/m^2}$. Such high luminance tends to induce an increase in power consumption due to the increased voltage and a reduction of lifetime due to the high current density.

By contrast, AM-OLEDs can solve these problems of PM-OLEDs due to the driving control using TFTs, but the fabrication process of TFTs increases the cost because TFTs require a long and complex fabrication process.

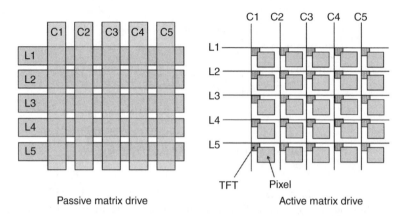

Figure 8.2   Passive-matrix OLED (PM-OLED) and active-matrix OLED (AM-OLED)

Table 8.1   Comparison of passive-matrix OLED (PM-OLED) and active-matrix OLED (AM-OLED)

Drive method	Passive-matrix	Active-matrix
TFT	None	More than two TFTs
Structure	Simple	Complex
Power consumption	Large	Small
Lifetime	Short	Long
Pixel number	Small	large

8.3   Passive-Matrix OLED Display

The device structures of PM-OLED displays are simple compared to AM-OLED displays because PM-OLED displays do not use an active-matrix array such as TFT. However, PM-OLED displays usually require an unique technology. This is cathode separator technology, not being utilized in static-drive OLED and AM-OLED displays. PM-OLED displays require multiple cathode lines on organic layers. Since organic layers in OLEDs should be protected from air and moisture, the cathode lines have to be fabricated in vacuum or inert gas environment. Therefore, common patterning techniques such as photolithography cannot be used. An alternative technology may be cathode metal deposition using a fine metal mask with a stripe pattern. However, the treatment of such a fine metal mask with stripe pattern is not very easy in production. Based on this background, the cathode separator technology was developed as a realistic production technology.

The cathode separator was reported by Pioneer [2]. Figure 8.3 shows schematic process flow using cathode separator technology. The cathode separator has a reverse tapered or T-shaped cross section, being formed prior to the deposition of organic layers and cathode metals. Owing to this structure, cathode metals on organic layers of each line are separated from each other. This technology was applied to the world's first commercialized OLED display by Pioneer in 1997.

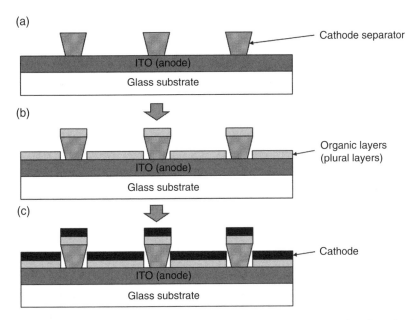

Figure 8.3 Schematic illustrations of the cathode micro-patterning processes: (a) forming of cathode separators; (b) evaporation of organic materials; (c) evaporation of cathode metal. [2] (Copyright 1997 The Japan Society of Applied Physics)

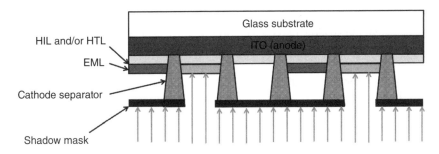

Figure 8.4 RGB patterning method using a precision shadow-mask and cathode separator

While Pioneer's commercialized OLED display (see Fig. 1.2) was a passive-matrix, monochrome, small molecule and evaporation-type OLED, various types of passive-matrix OLEDs have been developed and commercialized.

Kubota and Fukuda et al. of Pioneer reported a full-color passive-matrix OLED display, by developing a selective deposition of small molecule organic materials using a precision shadow-mask and high-accuracy mask moving mechanism [3]. One of the important key technologies is a gap control between the substrate and the mask. In their technology, it was reported that the distance between the substrate and the shadow-mask is maintained at 5 μm, using the cathode separators as stoppers for the shadow mask, as shown in Fig. 8.4. The cathode separator plays roles in not only separation of the cathode metal but also gap control between the substrate and the shadow mask. After one emission layer with a certain

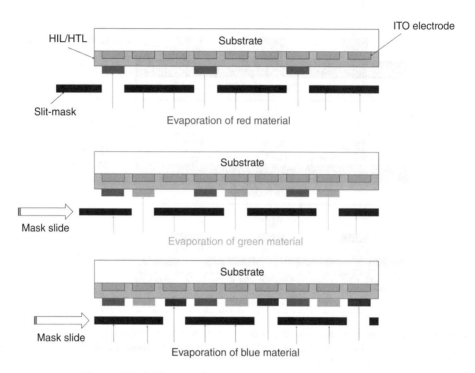

Figure 8.5   RGB patterning method using a shadow-mask [3, 4]

color is deposited, the shadow mask is slid, followed by the deposition of the next emission layer with a different color, as shown in Fig. 8.5. They also developed a high-accuracy shadow-mask moving mechanism in a vacuum chamber, because OLED devices should be kept in a vacuum condition during deposition of organic materials and cathode metal, in order to eliminate attack from air containing water and oxygen. They fabricated a 5.2″ diagonal full-color PM-OLED display with QVGA format (320×240 dots), luminance of 150 cd/m² and 64 gray levels for each color.

Mori et al. of NEC also reported a full-color passive-matrix OLED display by using the slit-mask sliding method [5]. They fabricated a 5.7″ diagonal full-color PM-OLED display with QVGA format (320×240 dots), luminance of 140 cd/m².

As described in the previous section, it is difficult to fabricate large size displays using PM-OLEDs. As an alternative technology, tiling has been developed. Mitsubishi Electric Corporation and Tohoku Pioneer Corporation collaboratively developed a tiling display technology using PM-OLEDs [6]. They fabricated a 155″ (3.93 m) diagonal prototype using 2880 PM-OLED panels, as shown in Fig. 8.6. The system can be applied to not only flat panel displays but also cubic displays such as a terrestrial globe as shown in Fig. 1.3.

8.4   Active-Matrix OLED Display

Active-matrix OLED (AM-OLED) displays can be applied to various displays such as mobile phones, smart phones, digital camera, TV, etc. AM-OLED displays usually require a TFT back-plane on which an OLED device is fabricated. Various TFT technologies such as amorphous-Si

Figure 8.6 A 155″ tiling OLED display systems using PM-OLEDs [6] (Photo: provided from Mitsubishi Electric Corporation)

(a-Si) TFT, low temperature poly-silicon (LTPS) TFT, oxide TFT, organic TFT (OTFT), etc. have been investigated and applied to OLED displays. This section describes TFT circuit technologies, TFT device technologies, and commercialized and prototype AM-OLED displays.

8.4.1   TFT Circuit Technologies

Active-matrix OLED displays require different TFT circuits from those used in LCDs because LCD is a capacitive device while OLED is a current-based device. Each pixel of AM-LCDs can usually be driven by one TFT, but each pixel of an AM-OLED has to be be driven by multiple TFTs.

The basic TFT circuits for AM-LCDs and AM-OLEDs are shown in Fig. 8.7. In AM-LCDs, the liquid crystal device is a capacitor, as shown in Fig. 8.7(a). The scanning lines (gate lines) are sequentially selected. That is, a turn-on signal is applied to the corresponding gate electrode of the transistors connecting them with the selected scanning line. In the turn-on period, a signal voltage corresponding to the required transmittance of the pixel of liquid crystal devices connecting with the data line (source line) is applied to the source electrode of the transistor through the data line. And then the signal voltage is applied to the liquid crystal device through the drain electrode of the transistor.

Figure 8.7(b) shows a simple two-transistor TFT circuit for AM-OLED. The TFT-1 connected with the scanning line (gate lines) is called the switching transistor and the TFT-2 connected with the OLED device is called driving transistor. As with the TFTs for AM-LCDs, the scanning lines are sequentially selected. A turn-on signal is applied to the corresponding gate electrode of the transistors connected to the selected scanning line. In the turn-on period, a signal voltage corresponding to the required luminance of the pixel of the OLED device is applied to the source electrode of the switching TFT (TFT-1) through the data line. Since the drain electrode of TFT-1 is connected to a capacitor and to the gate electrode of the driving transistor (TFT-2), the signal voltage is stored in the capacitor and

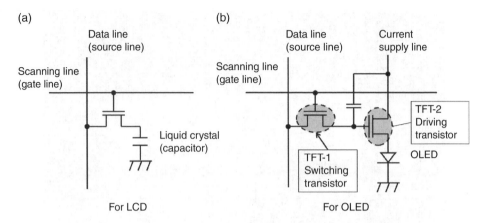

Figure 8.7   Basic TFT circuits for AM-LCDs and AM-OLEDs

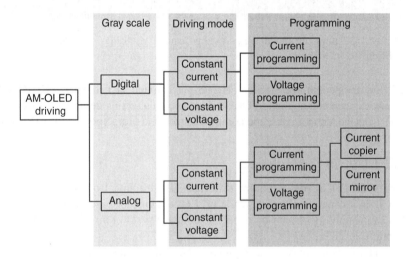

Figure 8.8   Classification of TFT driving circuits

applied to the gate electrode of TFT-2. The source electrode of TFT-2 is connected to the current supply line. Current from the supply line is supplied to each OLED device through TFT-2. The driving transistor (TFT-2) operates in its saturation region, where current is mainly dependent on the gate voltage.

The TFT circuit with two transistors is simple but has some problems. Since this simple circuit cannot compensate for the variation of threshold voltage, mobility, etc. of TFTs, fluctuation of TFT characteristics leads to non-uniformity of the picture. In order to compensate for variations in TFT characteristics, various types of TFT circuits have been studied, usually using 4–6 TFTs in each pixel.

The classification of these driving circuits is shown in Fig. 8.8, where first the driving methods are divided to digital and analog by gray scale methods.

Digital driving method is an easy way to control gray scale levels in a sense. In digital driving, temporal or spatial dither methods are used. In some cases, both dither methods are combined. In the spatial dither, pixel elements are divided into plural sub-pixels. Therefore, very small sub-pixel size is required, increasing the difficulty of the fabrication process. In temporal dither, emitted time is divided to multiple sub-periods. Therefore, very high frequency is required, increasing the difficulty of TFT driving.

Currently, major driving methods are analog driving, while the gray scale control is not so easy, due to the fluctuations of TFT characteristics. The main fluctuations of TFTs are caused by threshold voltage, mobility, and their changes under driving. For compensating such fluctuations, various compensating TFT circuits have been proposed and investigated, as described below.

The second classification term is the driving mode: constant current and constant voltage. Constant voltage is an easier method. However, it should be noted that current varies with changing temperature, while OLED luminance is determined by current. Therefore, currently, the main driving mode is constant current.

The third classification term is the programming method. One is voltage programming and the other is current programming. Voltage programming methods tend to be easier than current programming methods. In the voltage programming method, a data voltage is written to the gate of the driving TFT by means of a selection switch and is stored on the storage capacitor. Figure 8.9 shows an example of voltage a programming circuit reported by Dawson et al. of Sarnoff [7]. This circuit consists of four TFTs and two capacitors in each pixel.

Yoo et al. of Samsung SDI Co., Ltd also reported a voltage programming pixel circuit with six TFTs and one capacitor [8]. The circuit can compensate the V_{th} variation on the driving TFT in each pixel, while mobility variations remain unaffected.

One of the problems of the voltage programming methods is pixel-to-pixel non-uniformity in luminance which is induced by the mobility fluctuation of driving TFTs. Indeed, LTPS tends to have non-negligible mobility fluctuation.

The issue of non-uniformity in luminance in voltage programming methods can be solved by addressing the pixel circuit with current programming. These circuits sense the addressing current and then mirror the same current or a multiple of that current to the OLED device.

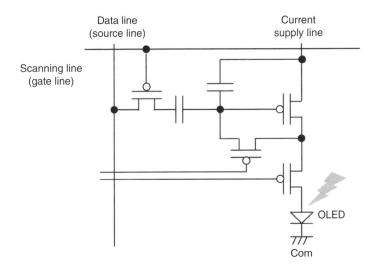

Figure 8.9   An example of a voltage programming pixel circuit [7]

In current programming methods, well-known methods are the current mirror [9] shown in Fig. 8.10 and the current copier [10] shown in Fig. 8.11. Each TFT circuit has to be programmed to supply the required current, which is determined by the picture images. Since the luminance of each pixel in OLED displays is proportional to the supplied current, the current programming method can give good uniformity. In other words, this method can automatically compensate for the variations of the mobility and threshold voltage. However, programming these circuits is slow due to the large parasitic column capacitance and the low program currents, particularly at low gray levels and for large displays.

One of the alternative approaches for solving picture non-uniformity is a utilization of IGZO(In-Ga-Zn-O)-TFTs, because they tend to have small distribution of TFT

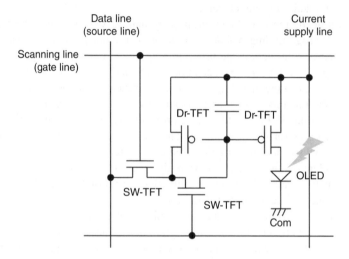

Figure 8.10   An example of current mirror TFT circuit [9]

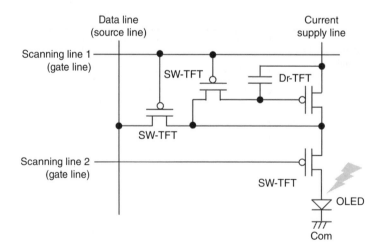

Figure 8.11   An example of current copier TFT circuit [10]

characteristics. Saito et al. ay Toshiba reported that a 3″ AM-OLED with IGZO TFTs showed excellent luminance uniformity, even though they utilized a simple non-compensating TFT circuit with two transistors and one capacitor [11].

8.4.2 TFT Device Technologies

OLED devices can be driven by various TFTs such as a-Si, poly-Si, micro-crystal Si, single crystal Si, oxide TFT, or organic TFT. They are classified as shown in Fig. 8.12, and the comparison of these TFTs is shown in Table 8.2.

Currently, the most commonly used TFTs for AM-OLED displays are low temperature poly-Si (LTPS) and IGZO (In-Ga-Zn-O), which is a kind of oxide TFT. Single crystal silicon is also used for AM-OLED displays, especially for micro-display applications. Although a-Si-TFTs can drive AM-OLED displays in spite of their low mobility, the a-Si-TFT still has the intrinsically serious problem of driving stability. Organic TFTs may be a technology for the future and will be described in Section 11.2.

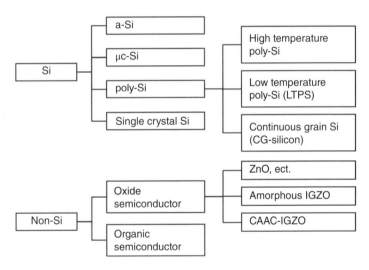

Figure 8.12 Classification of TFTs for OLED displays

Table 8.2 Comparison of TFTs for AM-OLED displays

	a-Si	LTPS[*]	IGZO[†]	Organic
Drive ability	Poor	Excellent	Excellent	Still poor
Field effect mobility (cm²/Vs)	0.3~1	~100	10~30	0.1~1
V_{th} uniformity	Good	Poor	Good	Issue
V_{th} stability	Poor	Excellent	Good	Poor

[*] including CG-silicon
[†] including CAAC-IGZO

8.4.2.1   LTPS

Low temperature poly-Si-TFT (LTPS-TFT) has the advantages of excellent mobility (around $100\,cm^2/Vs$) and good stability, even though the process cost is much higher than a-Si-TFT, IGZO-TFT, etc. Indeed, most commercial AM-OLED displays use LTPS-TFT.

Fabrication processes and equipment for LTPS-TFTs are significantly different from conventional amorphous Si (a-Si) TFTs. LTPS requires linear laser crystallization apparatus, dehydrogenation apparatus, and ion-doping apparatus, among other things.

LTPS can be applied to CMOS circuits which use complementary pairs of n-type and p-type transistors. Due to the high mobility, LTPS can achieve an integration of driver circuits around a display periphery. This is a big advantage in small and/or mobile displays because it significantly reduces the cost of the driver electronics. The LTPS fabrication process, based on the excimer-laser annealing (ELA) process, is now a mature technology through the production of AM-LCDs over many years. The current beam size for ELA process is 130 cm so that two scans can cover G8 glass [12].

Problems of LTPS-TFTs for OLEDs are non-uniformity of threshold voltage and mobility, due to variations of the TFT characteristics. Because of this non-uniformity of TFT performance, compensation TFT circuits with 4 ~ 6 TFTs in each pixel is widely used as mentioned in the previous section.

8.4.2.2   Oxide TFT, IGZO

Recently, oxide TFTs such as IGZO (In-Ga-Zn-O) TFTs have attracted much attention for backplanes for OLED displays because of the high mobility of around $10 \sim 30\,cm^2/Vs$, good uniformity, low leakage current, etc. The other advantage of oxide TFTs is relatively low fabrication cost, because oxide TFTs can be fabricated by modifying a-Si-TFT fabrication lines. Therefore, many prototype AM-OLED displays using IGZO-TFT have been developed and reported, as shown in Tables 8.4 and 8.5. Since oxide semiconductor materials such as IGZO are oxygen deficient, the active carriers are electrons. Therefore, oxide semiconductor can only realize NMOS.

In 2004, Nomura et al of Hosono's group at the Tokyo Institute of Technology proposed a transparent amorphous oxide semiconductor from the In-Ga-Zn-O system (a-IGZO) for the active channel in transparent thin-film transistors [13]. They deposited the a-IGZO on polyethylene terephthalate at room temperature and reported achieving the mobilities higher than $10\,cm^2/Vs$.

Various kinds of TFT structures have been investigated and reported, and typical TFT structures are shown in Fig. 8.13, where type (a) does not have an etch stopper (ES) while type (b) does. The etch-stopper (ES) structure shown in Fig. 8.13(b) is widely used because the etch stopper can protect the oxide semiconductor during the TFT process such as etching and plasma processes.

The amorphous IGZO film can be formed by a DC sputtering method at room temperature.

Yamazaki et al. of Semiconductor Energy Laboratories Co., Ltd (SEL) developed a c-axis aligned crystal (CAAC) IGZO [14, 15]. A CAAC oxide semiconductor (CAAC-OS) has a crystal structure with aligned orientation, and provides FETs with extremely low off-state current. The mobility is higher than $30\,cm^2/Vs$.

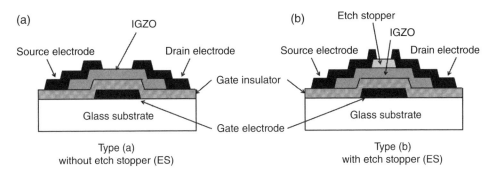

Figure 8.13   An example of the device structure of IGZO-TFT

Based on the promising potential and current developmental status of IGZO, IDTech predicted a \$16 billion market for OLED display modules driven by oxide TFT backplane [16].

8.4.2.3   a-Si

Although the field effect mobility of a-Si-TFT is as low as $0.3–1.0\,cm^2/Vs$, AM-OLED displays can be driven by a-Si-TFT. One of the merits of a-Si-TFT is uniformity of the threshold of TFT. Since a-Si-TFT is an established technology for AM-LCDs, the application of a-Si-TFT to AM-OLEDs has been investigated as shown in Table 8.3.

However, in the application of a-Si-TFT to AM-OLEDs, one of the serious issues is the large threshold shift under driving. Since OLEDs are current driven devices, the threshold shift under driving in a-Si-TFT for AM-OLEDs is much larger than that for AM-LCDs. Therefore, for driving OLEDs, the compensating circuit of a-Si-TFT is necessarily required for driving AM-OLEDs. For example, Tsujimura et al. at International Display Technology (Japan) proposed a voltage-controlled four-TFT compensation circuit [17].

While device and driving technologies for AM-OLED displays driven by a-Si-TFT have been developed, the development activity on AM-OLEDs with a-Si-TFT has been reduced recently. It seems to be recognized that a-Si-TFT is almost impossible to be applied to commercial AM-OLED products because of the unsolved problem of the large threshold shift under driving of a-Si-TFT.

8.4.3   *Commercialized and Prototype AM-OLED Displays*

Practical full-color AM-OLED displays can be fabricated by combining TFT technologies and full-color OLED technologies.

Bottom emitting device structures tend to have a restriction of aperture ratio as described in Section 5.1. Generally, top emitting device structures tend to show larger aperture ratio than bottom emitting ones, enabling higher luminance and longer lifetime, while the fabrication process is more complicated than bottom emitting. Typical device structure of top emitting full-color AM-OLED displays is illustrated in Fig. 8.14. In particular, for high resolution mobile display, we tends to necessarily adopt the top emitting device structure from the aperture ratio point of view.

Table 8.3 Examples of AM-OLED displays driven by a-Si-TFTs

Affiliation[*]	Year	Size (inch diagonal)	Format[†]	Full-color method	Emitting direction	OLED Process	Comment	Ref.
IDT	2003	20	WXGA/HDTV	RGB side-by-side	Top	Evaporation		[17]
AU Optronics	2003	4	234×160	RGB side-by-side	Bottom	Evaporation		[18]
Casio	2004	2.1	160×128	RGB side-by-side	Bottom	Ink-jet	101 ppi	[19]
Samsung	2005	7.0	HVGA	RGB side-by-side	Bottom	Ink-jet		[20]
LG Chem	2008	3.5	QVGA	RGB side-by-side	Top	Evaporation	Inverted	[21]

[*]Affiliation of the first author.
[†]QVGA: 320×240, HVGA:480×320.

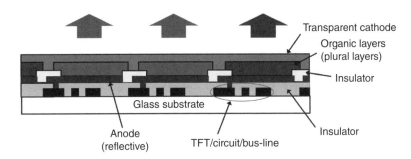

Figure 8.14   Device structures of typical top emitting full-color AM-OLED displays

For full-color technologies, either RGB side-by-side or white/CF tend to be used for practical AM-OLED displays. The merits and demerits of these two methods are described in Section 5.4. At present, for very high resolution mobile displays and for very large size displays the white/CF method tends to be adopted because it can avoid the difficulty of the fine metal mask evaporation.

Table 8.4 summarizes several typical commercialized AM-OLED displays.

The world's first AM-OLED display was commercialized in 2003 by SK Display, which is a joint company of Sanyo Electric and Eastman Kodak [22]. This was a mobile display and was applied to digital still cameras. In 2007, Samsung commercialized the world's first AM-OLED displays for mobile phones [23]. One of the recent trends in mobile displays is high resolution. For example, Samsung commercialized an 8.4″ display with 2560×1600 dots.

On the other hand, as OLED-TVs, Sony commercialized the world's first OLED-TV in 2007 [24]. The display size was 11″. Recently, Sony commercialized large AM-OLED displays with 25–30″ for broadcast monitors [25]. In addition, LG Display recently commercialized large OLED-TVs of 55″. One of them has a curved shape.

Table 8.5 summarizes several typical prototype AM-OLED displays, while prototype AM-OLED displays fabricated using ink-jet printing, a-Si-TFT, or flexible substrate are summarized in Tables 6.2, 8.3, and 10.3, respectively.

At the early R&D stages of AM-OLED displays, the display that had the biggest impact was Sony's 13″ prototype AM-OLED presented in 2001 [9]. This was the largest OLED display at that time and generated beautiful images, adopting several cutting-edge technologies such as top emission device structure with micro-cavity effect, and combination with color filter and solid state encapsulation. These technologies were applied to the world-first OLED-TV commercialized in 2007 [24].

While this type of AM-OLED display was fabricated by using LTPS and vacuum evaporation processes with fine metal mask, many prototype AM-OLED displays with oxide TFTs and/or white/CF structure have been developed and reported. The developmental trends were high resolution for mobile displays and large size for TVs.

As a mobile display, Semiconductor Energy Laboratory (SEL) in Japan developed a 13.3″ AM-OLED display with 8K4K format (7680×4320 dots) with 664 ppi [36] (see Fig. 1.7). Moreover, SEL developed a very high resolution AM-OLED display, which was a 2.8″ display with 1058 ppi [43] (Fig. 8.15).

Table 8.4 Several commercialized AM-OLED displays

Affiliation	Year	Size	Format	TFT	RGB	Emission	OLED Process	Comment	Ref.
SK Display	2003	2.16	521×218	LTPS	Side-by-side	Bottom	Evaporation		[22]
Samsung	2007	2.4	QVGA	LTPS	Side-by-side	Top	Evaporation		[23]
Sony	2007	11	960×540	LTPS	Side-by-side	Top	Evaporation	With CF	[24]
Sony	2011	25	1920×1080 (FULL HD)	LTPS	Side-by-side	Top	Evaporation	With CF	[25]
LG Display	2013	55"	FHD	IGZO	White/CF(RGBW)	Bottom	Evaporation	Flat & Curved	[26]
Samsung	2014	8.4	2560×1600	LTPS	Side-by-side	Top	Evaporation	Pentile matrix sub-pixel design	
Sony	2014	30	4096×2160 (4K 2K)	LTPS	Side-by-side	Top	Evaporation	With CF	[27]
LG Display	2015	55	3840×2160		White/CF(RGBW)		Evaporation	Curved	

Table 8.5 Several prototypes of AM-OLED displays

Affiliation[a]	Year	Size	Format[b]	TFT[c]	RGB	Emission	OLED Process	Comment	Ref.
Sony	2001	13	800×600	LTPS	RGB side-by-side	Top	Evaporation	With CF	[9]
Sanyo Electric	2003	2.5	320×240 (QVGA)	LTPS	White/CF	Bottom	Evaporation		[28]
Sony	2004	12.5	854×480	LTPS	White/CF	Top	Evaporation		[29]
Samsung SDI	2005	2.6	640×480	LTPS	RGB side-by-side	Top	Evaporation	302 ppi	[30]
Sony	2007	27	1920×1080	Micro-Si	White/CF	Top	Evaporation		[31]
LG	2007	3.5	220×176	IGZO	RGB side-by-side	Top	Evaporation		[32]
SEL	2012	13.5	4K2K	CAAC-IGZO	White/CF	Top	Evaporation	326 ppi	[33]
AUO	2012	32	1920×1080	IGZO	RGB stripe	Bottom	Evaporation		[34]
Sony	2013	56	4K2K	IGZO	White/CF	Top	Evaporation		[35]
SEL	2014	13.3	8K4K	CAAC-IGZO	White/CF	Top	Evaporation	664 ppi	[36]
SCSOT[d]	2014	31	FHD	IGZO	FMM	Bottom	Evaporation		[37]
Samsung	2014	55	FHD	LTPS	FMM	Bottom	Evaporation		[38]
Panasonic	2014	55	4K2K	IGZO	Printing	Top	Printing		[39]
BOE	2014	55	4K2K		White/CF	Bottom	Evaporation		[40]
AUO	2014	65	FHD	IGZO	FMM	Bottom	Evaporation		[41]
LG	2014	77	4K2K	Oxide TFT	White/CF	Bottom	Evaporation	Curved	[42]
SEL	2015	2.78	WQHD (2560×1440)	CAAC-IGZO	White/CF	Top	Evaporation	1058 ppi	[43]

[a] Affiliation of the first author.
SEL: Semiconductor Energy Laboratory Co., Ltd.
[b] FHD: 1920×1080, 4K2K: 3840×2160, 8K4K: 7680×4320.
[c] CAAC: c-axis aligned crystal.
[d] Shenzhen China Star Optoelectronics Technology Co., Ltd.

Figure 8.15 An ultra high definition AM-OLED developed by Semiconductor Energy Laboratory
(SEL) [43]. (provided by Semiconductor Energy Laboratory)
Display size: 2.8″ (61 mm × 35 mm)
Number of pixels: 2560 × 1440 (WQHD)
Resolution: 1058 ppi
Color: Full color
Device structure: White tandem OLED (top emission) + color filter
Driving: Active-matrix with CAAC-IGZO TFT

As OLED-TVs, Sony [35], Panasonic [39], Samsung [38], LG Display [42], AUO [42], and
BOE [40] all developed 55″ or 56″ OLED-TVs in 2012–2014. In addition, AUO [41] and LG
Display [42] developed 65″ and 77″ OLED-TV, respectively.

AM-OLED technologies can also be applied to micro-displays, which can be used, for
example, for head-mounted displays. For example, eMagin Corporation developed active-
matrix OLED micro-displays [44]. They fabricated an OLED on a silicon wafer with
driving circuits. Since the silicon wafer is not transparent, the OLED has to be the top
emitting type. In order to obtain full colors, the white emission OLED is combined with a
color filter.

References

[1] T. Kurita, T. Kondo, *Proc. IDW'00*, LCT4-1 (p. 69) (2000).
[2] K. Nagayama, T. Yahagi, H. Nakada, T. Tohma, T. Watanabe, K. Yoshida, S. Miyaguchi, *Jpn. J. Appl. Phys.*,
    36(11B), L1555–L1557 (1997).

[3] H. Kubota, S. Miyaguchi, S. Ishizuka, T. Wakimoto, J. Funaki, Y. Fukuda, T. Watanabe, H. Ochi, T. Sakamoto, T. Miyake, M. Tsuchida, I. Ohshita, T. Tohma, *Journal of Luminescence*, **87**, 56–60 (2000);Y. Fukuda, T. Watanabe, T. Wakimoto, S. Miyaguchi, M. Tsuchida, *Synthetic Metals*, **111–112**, 1–6 (2000).

[4] I. Ohshita, *PIONEER R&D*, **22**, 24–33 (2013).

[5] K. Mori, Y. Sakaguchi, Y. Iketsu, J. Suzuki, *Displays*, **22**, 43–47 (2001).

[6] Z. Hara, K. Maeshima, N. Terazaki, S. Kiridoshi, T. Kurata, T. Okumura, Y. Suehiro, T. Yuki, *SID 10 Digest*, 25.3 (2010);S. Kiridoshi, Z. Hara, M. Moribe, T. Ochiai, T. Okumura, *Mitsubishi Electric Corporation Advance Magazine*, **45**, 357 (2012).

[7] R. M. A. Dawson, M. G. Kane, *SID 01 Digest*, 24.1 (p. 372) (2001).

[8] K.-J. Yoo, S.-H. Lee, A.-S. Lee, C.-Y. Im, T.-M. Kang, W.-J. Lee, S.-T. Lee, H.-D. Kim, H.-K. Chung, *SID 05 Digest*, 38.2 (p. 1344) (2005).

[9] T. Sasaoka, M. Sekiya, A. Yumoto, J. Yamada, T. Hirano, Y. Iwase, T. Yamada, T. Ishibashi, T. Mori, M. Asano, S. Tamura, T. Urabe, et al., *SID 01 Digest*, 24.4 L (p. 384) (2001); J. Yamada, T. Hirano, Y. Iwase, T. Sasaoka, *Proc. AM-LCD'02*, D-2 (p. 77) (2002).

[10] M. Ohta, H. Tsutsu, H. Takahara, I. Kobayashi, T. Uemura, Y. Takubo, *SID 03 Digest*, 9.4 (p. 108) (2003).

[11] N. Saito, T. Ueda, S. Nakano, Y. Hara, K. Miura, H. Yamaguchi, I. Amemiya, A. Ishida, Y. Matsuura, A. Sasaki, J. Tonotani, M. Ikagawa, *Proc. IDW'10*, AMD-9 (2010).

[12] M. Mativenga, D. Geng, J. Jang, *SID 2014 Digest*, 3.1 (p. 1) (2014).

[13] K. Nomura, H. Ohta, A. Takagi, T. Kamiya, M. Hirano, H. Hosono, *Nature*, **432**, 488–492 (2004).

[14] S. Yamazaki, J. Koyama, Y. Yamamoto, K. Okamoto, *SID 2012 Digest*, 15.1 (p. 183) (2012).

[15] S. Yamazaki, *SID 2014 Digest*, 3.3 (p. 9) (2014).

[16] K. Ghaffarzadeh, IDTechEX, "Metal Oxide TFT Backplanes for Displays 2014–2024: Technologies, Forecasts, Players" (2014).

[17] T. Tsujimura, Y. Kobayashi, K. Murayama, A. Tanaka, M. Morooka, E. Fukumoto, H. Fujimoto, J. Sekine, K. Kanoh, K. Takeda, K. Miwa, M. Asano, N. Ikeda, S. Kohara, S. Ono, C.-T. Chung, R.-M. Chen, J.-W. Chung, C.-W. Huang, H.-R. Guo, C.-C. Yang, C.-C. Hsu, H.-J. Huang, W. Riess, H. Riel, S. Karg, T. Beierlein, D. Gundlach, F. Libsch, M. Mastro, R. Polastre, A. Lien, J. Sanford, R. Kaufman, *SID 03 Digest*, 4.1 (p. 6) (2003); S. Ono, Y. Kobayashi, K. Miwa, T. Tsujimura, *Proc. IDW'03*, AMD3/OEL4-2 (p. 255) (2003).

[18] J.-J. Lih, C.-F. Sung, M. S. Weaver, M. Hack, J. J. Brown, *SID 03 Digest*, 4.3 (p. 14) (2003).

[19] T. Shirasaki, T. Ozaki, K. Sato, M. Kumagai, M. Takei, T. Toyama, S. Shimoda, T. Tano, *SID 04 Digest*, 57.4 L (p. 1516) (2004).

[20] D. Lee, J.-K. Chung, J.-S. Rhee, J.-P. Wang, S.-M. Hong, B.-R. Choi, S.-W.Cha, N.-D.Kim, K. Chung, H. Gregory, P. Lyon, C. Creighton, J. Carter, M. Hatcher, O. Bassett, M. Richardson, P. Jerram, *SID 05 Digest*, 527 (2005).

[21] J. K. Noh, M. S. Kang, J. S. Kim, J. H. Lee, Y. H. Ham, J. B. Kim, M. K. Joo, S. Son, *Proc. IDW'08*, OLED3-1 (p. 161) (2008).

[22] K. Mameno, R. Nishikawa, K. Suzuki, S. Matsumoto, T. Yamaguchi, K. Yoneda, Y. Hamada, H. Kanno, Y. Nishio, H. Matsuola, Y. Saito, S. Oima, N. Mori, G. Rajeswaran, S. Mizukoshi, T. K. Hatwar, *Proc. IDW'02*, **235** (2002).

[23] News release of KDDI, 20 March 2007. www.kddi.com/corporate/news_release/2007/0320/

[24] News release of Sony Corporation, 1 October 2007. www.sony.jp/CorporateCruise/Press/200710/07-1001/

[25] News release of Sony Corporation, 9 September 2011. www.sony.co.jp/SonyInfo/News/Press/201109/11-107/

[26] C.-W. Han, J.-S. Park, Y.-H. Shin, M.-J. Lim, B.-C. Kim, Y.-H. Tak, B.-C. Ahn, *SID 2014 Digest*, 53.2 (p. 770) (2014); J.-S. Yoon, S.-J. Hong, J.-H. Kim, D.-H Kim, T. Ryosuke, W.-J.Nam, B.-C. Song, J.-M. Kim, P.-Y. Kim, K.-H. Park, C.-H. Oh, B.-C. Ahn, *SID 2014 Digest*, 58.2 (p. 849) (2014).

[27] News release of Sony Corporation, 9 September 2011. www.sony.co.jp/SonyInfo/News/Press/201411/14-114/

[28] K. Mameno, S. Matsumoto, R. Kishikawa, T. Sasatani, K. Suzuki, T. Yamaguchi, K. Yoneda, Y. Hamada, N. Saito, *Proc. IWD'03*, AMD4/OLED5-1 (p. 267) (2003).

[29] M. Kashiwabara, K. Hanawa, R. Asaki, I. Kobori, R. Matsuura, H. Yamada, T. Yamamoto, A. Ozawa, Y. Sato, S. Terada, J. Yamada, T. Sasaoka, S. Tamura and T. Urabe, *SID 04 Digest*, 29.5 L (p. 1017) (2004).

[30] K.-J. Yoo, S.-H. Lee, A.-S. Lee, C.-Y. Im, T.-M. Kang, W.-J. Lee, S.-T. Lee, H.-D. Kim, H.-K. Chung, *SID 05 Digest*, 38.2 (p. 1344) (2005).

[31] T. Urabe, T. Sasaoka, K. Tatsuki, J. Takai, *SID 07 Digest*, 13.1 (p. 161) (2007); T. Arai, N. Morosawa, Y. Hiromasu, K. Hidaka, T. Nakayama, A. Makita, M. Toyota, N. Hayashi, Y. Yoshimura, A. Sato, K. Namekawa, Y. Inagaki, N. Umezu, K. Tatsuki, *SID 07 Digest*, 41.2 (p. 1370) (2007).

[32] H.-N. Lee, J. Kyung, S. K. Kang, D. Y. Kim, M.-C. Sung, S.-J. Kim, C. N. Kim, H. G. Kim, S.-t. Kim, *SID 07 Digest*, 68.2 (p. 1826) (2007).

[33] S. Yamazaki, J. Koyama, Y. Yamamoto, K. Okamoto, *SID 2012 Digest*, 15.1 (p. 183) (2012);T. Tanabe, S. Amano, H. Miyake, A. Suzuki, R. Komatsu, J. Koyama, S. Yamazaki, K. Okazaki, M. Katayama, H. Matsukizono, Y. Kanzaki, T. Matsuo, *SID 2012 Digest*, 9.2 (p. 88) (2012);S. Eguchi, H. Shinoda, T. Isa, H. Miyake, S. Kawashima, M. Takahashi, Y. Hirakata, S. Yamazaki, M. Katayama, K. Okazaki, A. Nakamura, K. Kikuchi, M. Niboshi, Y. Tsukamoto, S. Mitsui, *SID 2012 Digest*, 27.4 (p. 367) (2012).

[34] T.-H. Shih, T.-T. Tsai, K.-C. Chen, Y.-C. Lee, S.-W. Fang, J.-Y. Lee, W.-J. Hsieh, S.-H. Tseng, Y.-M. Chiang, W.-H. Wu, S.-C. Wang, H.-H. Lu, L.-H. Chang, L. Tsai, C.-Y. Chen, Y.-H. Lin, *SID 2012 Digest*, 9.3 (p. 92) (2012).

[35] News release of Sony Corporation, 8 January 2013. www.sony.co.jp/SonyInfo/News/Press/201301/13-002/index.html

[36] S. Yamazaki, *SID 2014 Digest*, 3.3 (p. 9) (2014); S. Kawashima, S. Inoue, M. Shiokawa, A. Suzuki, S. Eguchi, Y. Hirakata, J. Koyama, S. Yamazaki, T. Sato, T. Shigenobu, Y. Ohta, S. Mitsui, N. Ueda, T. Matsuo, *SID 2014 Digest*, 44.1 (p. 627) (2014).

[37] C. Y. Su, W.-H. Li, L.-Q. Shi, X.-W. Lv, K.-Y. Ko, Y.-W. Liu, J.-C. Li, S.-C. Liu, C.-Y. Tseng, Y.-F. Wang, C.-C. Lo, *SID 2014 Digest*, 58.1 (p. 846) (2014).

[38] M. Choi, S. Kim, J.-m. Huh, C. Kim, H. Nam, *SID 2014 Digest*, 3.4L (p. 13) (2014).

[39] H. Hayashi, Y. Nakazaki, T. Izumi, A. Sasaki, T. Nakamura, E. Takeda, T. Saito, M. Goto, H. Takezawa, *SID 2014 Digest*, 58.3 (p. 853) (2014).

[40] Exhibition at SID 2014 Display Week (2014).

[41] T.-H. Shih, H.-C. Ting, P.-L. Lin, C.-L. Chen, L. Tsai, C.-Y. Chen, L.-F. Lin, C.-H. Liu, C.-C. Chen, H.-S. Lin, L.-H. Chang, Y.-H. Lin, H.-J. Hong, *SID 2014 Digest*, 53.1 (p. 766) (2014).

[42] C.-W. Han, J.-S. Park, Y.-H. Shin, M.-J. Lim, B.-C. Kim, Y.-H. Tak, B.-C. Ahn, *SID 2014 Digest*, 53.2 (p. 770) (2014).

[43] Demonstrated by Semiconductor Energy Laboratory (SEL) in Display Innovation 2014 (October, 2014); K. Yokoyama, S. Hirasa, N. Miyairi, Y. Jimbo, K. Toyotaka, M. Kaneyasu, H. Miyake, Y. Hirakata, S. Yamazaki, M. Nakada, T. Sato, N. Goto, *SID 2015 Digest*, 70.4 (p. 1039) (2015).

[44] O. Prache, *Displays*, 22, 49–56 (2001).

9

OLED Lighting

Summary

Lighting technologies have evolved from fire in the ancient world to electronic equipment in the present age. From the use of incandescent and fluorescent lamps there has been the recent major change to using light emitting diodes (LEDs), and now OLEDs for commercial lighting products. OLED lighting has several advantages such as it being thin, flat, lightweight, high color rendering index, potentially highly efficient, potentially flexible, no harmful materials (Hg, etc.), no UV emission, and weak blue emission.

This chapter describes OLED lighting technologies and applications, including their comparison with other forms of lighting.

Key words

lighting, out-coupling, white, multi-photon

9.1 Appearance of OLED Lighting

In human history, lighting technologies have evolved in stages. In the prehistoric world, light could only be obtained by using fires of wood, grass, and twigs. Then in the ancient age and the middle ages, light from oil and candles was added. So in a sense, for most of the history of humankind, the Sun and fire have been the only light source.

In the modern age, new light sources have appeared: gas lighting, incandescent lamps, and fluorescent lamps were invented. While incandescent lamps and fluorescent lamps have been widely used during our lifetime, the rapid generational change from incandescent and fluorescent lamps to light emitting diodes (LED) has occurred recently, due to features of LEDs such as high efficiency, long lifetime, and non-use of mercury.

OLED Displays and Lighting, First Edition. Mitsuhiro Koden.
© 2017 John Wiley & Sons, Ltd. Published 2017 by John Wiley & Sons, Ltd.

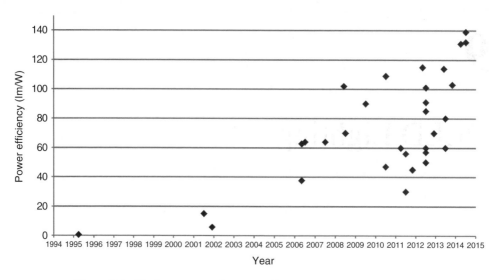

Figure 9.1   The improvement in power efficiency of white OLED devices

On the other hand, OLED lighting was first commercialized in 2011 by Lumiotec Inc., an OLED lighting venture company invested in by Mitsubishi Heavy Industry Ltd, Rohm Co., Ltd, Toppan Printing Co., Ltd, Mitsui & Co. Ltd, and Professor Kido [1].

The potential of OLED lighting was pointed out in early stages of OLED research and development. Kido et al. produced a scientific paper describing a white emitting OLED devices in 1994 [2] and in 1995 Kido et al. also wrote some significant phrases in their paper describing multi-stacked white emission OLED devices [3]. They wrote that the applications of white OLED devices could include paper-thin light sources, which are particularly useful in places that require a lightweight illumination device, such as in aircraft and space shuttles.

After the papers by Kido et al., white OLED technologies and OLED lighting technologies were actively developed, aimed at lighting applications. Accompanying the research and development of high efficiency phosphorescent materials, the efficiency of white OLEDs has been improved drastically over the past 20 years. Figure 9.1 shows the improvement in power efficiency of white OLED devices. In the development of OLED lighting, three significant technologies greatly contributed to the improvement in performance of white OLED devices. They are phosphorescent OLEDs, multi-photon technology, and out-coupling technologies.

9.2   Features of OLED Lighting

OLED lighting is an emerging solid-state lighting technology. It has several advantages such as being thin, flat, lightweight, high color rendering index, highly efficiency, design flexibility, potential for flexible devices, no harmful materials (Hg, etc.), no UV emission, and weak blue emission.

A comparison of the various forms of lighting is given in Table 9.1.

The incandescent lamp, which was invented by Thomas Edison in the 19th century, has been used for over 100 years. It can provide a warm white emission and is cheap to produce at around 1 dollar/klm. However, it has such two demerits; low efficiency of about 15 lm/W and short

Table 9.1 Comparison of forms of lighting

	Incandescent lamp	Fluorescent lamp	LED	OLED	
				Commercial level	Development level
Mechanism	Radiation by Joule heat	Inorganic phosphor excited by plasma	Inorganic semiconductor	Organic semiconductor	
Shape	Ball	Tube	Point	Planar	
Efficiency	~15 lm/W	~80 lm/W	~140 lm/W	~80 lm/W	~130 lm/W
Lifetime	1000~3000 hours	6000~12,000 hours	~40,000 hours	10,000~30,000 hours	~40,000 hours
cost	1 dollar/klm	1~3 dollars/klm	~5 dollars/klm	Higher than 100 dollars/klm	
Hg	None	Present	None	None	
UV light	None	Present	Present	None	
Blue light problem	None	None	Present	None	

lifetime. In particular, it is widely accepted that the low efficiency of incandescent lamps is not suitable for reducing greenhouse gas (GHG) emission. Based on this consensus, several governments have made laws prohibiting the use of incandescent lamps. Many manufacturers have recently stopped – or announced their intention to stop – manufacturing incandescent lamps.

Fluorescent lamps have been used since the middle of the 20th century and are still widely used at present. The efficiency of fluorescent lamps is about 80 lm/W, much higher than that of incandescent lamps. The lifetime of fluorescent lamps is about 6000 ~ 12,000 hours, again much longer than incandescent lamps. The cost is low, at about 1 ~ 3 dollars/klm. Due to these attractive features, in the latter half of the 20th century, most major lighting used fluorescent lamps. However, these lamps have some intrinsic issues such as that they contain the harmful element, mercury, and emit ultraviolet light. Currently, fluorescent lamps are being rapidly replaced by LEDs due to recent technological progress and the cost reduction of LEDs.

The most attractive features of LED seem to be long lifetime such as 40,000 hours and high efficiency such as 140 lm/W. In addition, since LEDs do not include harmful material such as Hg, they seem to be better than fluorescent lamp from an environment point of view.

Comparing with these three lighting technologies, it can be seen that OLEDs have intrinsically attractive features. One of the unique features of OLED lighting is its planar shape, while incandescent and fluorescent lamps have very much a three-dimensional shape, while LEDs are point-sources. Owing to this unique feature, OLEDs can realize planar lighting with low thickness and light weight. LEDs can also, of course, realize planar lighting by arranging many LED chips on flat substrates, but the following three issues should be borne in mind.

1. heat generation
2. greater thickness of planar lightings
3. energy use in constructing planar lightings

Heat generation is one of the issues of LEDs due to the intrinsic feature that LEDs are point sources. Because of the heat generation problem, LEDs require additional heat sink technologies. OLEDs, however, do not require any additional heat sink technologies, since they are intrinsically planar.

In order to construct planar lighting using LED chips, various plates are required. In addition to the substrate on which the LED chips are arranged, a light diffraction plate and heat sink are usually required. These substrates have certain thickness. On the other hand, an OLED device can be used as planar lighting by itself. Therefore, OLED lightings are much thinner than planar lighting using LED chips.

Then the power efficiency of LED planar lighting is much lower than that of individual LED chips, while the power efficiency of planar OLED lighting is almost the same as individual OLED lighting devices. In practice, the power efficiency of LED planar lighting is about 50 ~ 70% that of LED chips.

OLED lighting has no harmful material such as Hg, no UV-emission and weak blue emission. While the issue of blue light is debatable, it is a fact that there are arguments that blue light can damage human eyes and affect human bio-rhythms.

From these attractive features, it can be said that OLED lighting has high potential as the next generation of planar lighting. The possible market segments for OLED lighting are general lightings for houses, offices, shops, hospitals, personal equipment, outdoor, industrial, design lighting, as well as automotive lighting [4].

Currently, only small numbers of OLED lighting devices have penetrated the markets due to the fact that the cost is still too high and the performance of commercial products is still not good enough to compete with other forms of lighting.

However, the performance of OLED lighting is improving steadily. Table 9.1 shows current performances of OLED lighting in terms of commercial products and development levels. Looking at development levels, it can be seen that OLED lighting has great potential. In addition, it can be said that the cost can be reduced by the increase in production batch size and fabrication technology improvements and/or innovations.

The required specifications for OLED lighting should be considered in terms of the business competition with other lighting such as incandescent lamps, fluorescent lamps, and LEDs.

The primary requirements for OLED lightings are summarized in Table 9.2.

OLED luminance needs to be about $5000 \sim 7000 \, cd/m^2$, and it is not necessarily true that the higher, the better. Luminance as high as $10,000 \, cd/m^2$ is too glaring for human eyes. For a room with an area of about $15 \, m^2$, about $6000 \, lm$ are required. To obtaine this with OLED lighting panels with a luminance of $5000 \, cd/m^2$, the required panel area is only about $60 \, cm$ square.

Looking at recent progress and future forecasts of power efficiency of LEDs, the power efficiency of OLEDs needs to be higher than $150 \, lm/W$. This high efficiency is required not only for economic reasons but also because of the greenhouse gas issue. To reduce global CO_2 emission, high efficiency lighting is essential because lighting represents about 20% of global energy consumption. Since the efficiency of LEDs is predicted to be higher than $200 \, lm/W$ by 2020 [5], OLED lighting needs to achieve higher efficiency than $200 \, lm/W$. Moreover, high color rendering index (CRI) higher than about 90, and long lifetime such as 40,000 hours are also required for OLED lightings.

The cost of OLED lighting is still too high. The biggest reason is the currently small mass-production size, although there are other reasons such as expensive fabrication equipment and expensive materials – not only the OLED material itself but also such as glass substrates and ITO. The history of LCDs teaches us that major price decreases can occur with much higher speed than one could imagine, by increasing the substrate size and by reducing the costs of various materials and equipment. By expanding the production size, the cost of OLED lighting is expected to be drastically reduced, as the market for OLED lighting grows.

When the cost of OLED lighting is drastically reduced, it can compete with LEDs in business, especially in the field of planar large size lighting. Of course, one can fabricate large planar size lighting devices using LED chips, but planar lighting with LEDs has several disadvantages, related to the intrinsic fact that LEDs are point source devices. Therefore, for making planar lighting, LEDs require a diffuser plate, which reduces the efficiency by about 20–30%, and it increases the thickness. By contrast, OLEDs are intrinsically planar, and there is no requirement for a diffuser plate, so the efficiency of OLED planar lighting devices is

Table 9.2 The primary requirements for OLED lighting

Luminance	$5000 \sim 7000 \, cd/m^2$
Efficiency	$>150 \, lm/W$
Lifetime	$>40,000 \, hours$
CRI	>90

almost equal to that of the bare OLED devices, while the efficiency of LED planar lighting is lower than that of LED chips. In addition, OLED planar lighting can realize thinner and lighter units than LED planar lightings. Moreover, OLEDs can be used to produce flexible planar lighting devices, not easily possible with LEDs. As a conclusion, once the cost of OLEDs can be compete with LEDs and flexible OLED lighting is achieved, OLED lighting will be a major technology for planar lighting instead of LED planar lightings.

9.3   Fundamental Technologies of OLED Lighting

The fundamental technologies of OLED lighting are based on white OLEDs, which are described in Section 5.3, because lighting usually requires white emission.

While there are several methods for obtaining white emission, the most popular method is stacking a number of emission layers with different emission spectra. Figure 9.2 shows examples of white OLEDs with stacked emission layers, which usually contain two or three emission layers. With only two emission layers, a blue emission layer and a yellow or orange emission layer are usually combined. This technology is simpler than three stacking layers but it is difficult to achieve a high color rendering index using only two colors, but three stacking layers can realize a high color rendering index, using emission layers of red, green, and blue. This three-stack structure of emission layers can realize a color rendering index as high as 90 or more.

For fabricating white OLED lighting, the multi-photon technologies are very useful as described in Section 5.6. Multi-photon technologies can provide not only high efficiency and long lifetime but also an increase in production yield due to the increased total thickness of the organic layers. There are several types of white OLEDs with multi-photon technologies. Figure 9.3 shows three examples of white OLEDs with multi-photon structure. Example (a) combines two emission units, where each emission unit gives white emission. Example (b)

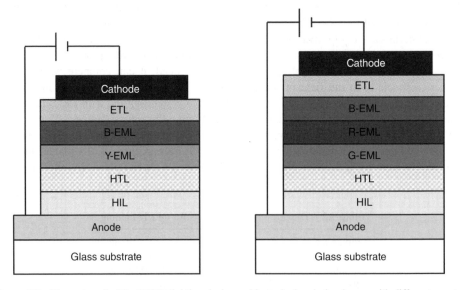

Figure 9.2   Examples of white OLED lighting devices with stacked emission layers with different spectra

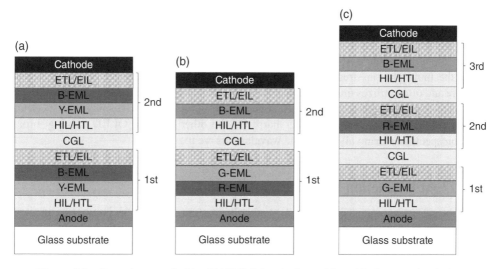

Figure 9.3 Several types of white OLED lighting devices with multi-photon technologies

	OLED display	OLED lighting
Substrate	TFT substrate	Glass substrate
Emission layer	RGB pixels	RGB stacking (white)
Device structure	Top emission	Bottom emission Multi-photon
Driving	Complex	Simple
Spectrum	Sharp for RGB	Broad
Competitors	LCD	LED
	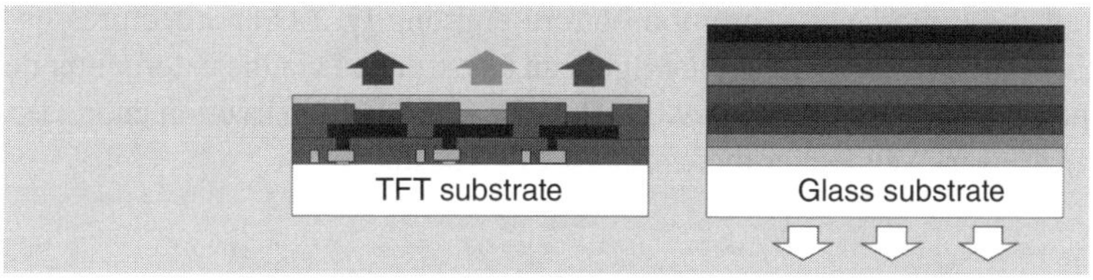 TFT substrate	Glass substrate

Figure 9.4 Comparison of typical AM-OLEDs and OLED lighting devices with multi-photon structure

also combines two emission units but these two emission units give different colors. As an actual technology, structure (b) is often applied to a structure where the first emission unit consists of red and green phosphorescent materials and the second emission unit consists of blue fluorescent material. This is due to the fact that red and green phosphorescent materials have commercial level performance but blue phosphorescent still has issues. Example (c) combines three emission units, where each unit gives each color. Of course, other multi-photon structures are possible to be fabricated.

A comparison of OLED displays and OLED lightings is summarized in Fig. 9.4. While both of them have various variations, as a typical case, Fig. 9.4 compares top emitting AM-OLED displays and white OLED lightings with the multi-photon structure.

While the device structure of AM-OLED displays tends to be complex due to the TFT structures for matrix drive and RGB pixels for full color images, etc., the device structure of OLED lighting is relatively simple because there is no requirement for full color picture images. In addition, while AM-OLEDs often use a top emitting device structure and scarcely utilize the multi-photon structure, OLED lighting usually adopts a bottom emitting device structure and often utilizes multi-photon device structure. Driving technologies for AM-OLED displays are more complicated than OLED lighting. In full-color OLED displays, the spectrum of each color, RGB, is required to be sharp. On the other hand, in OLED lighting, broad emission spectra covering all the visible wavelengths are required.

9.4   Light Extraction Enhancement Technologies

The light extraction enhancement (LEE) technologies (out-coupling technologies) are very important in OLED lighting because most of the emission in an OLED does not transmit if additional light extraction enhancement technologies are not applied. Indeed, while the internal quantum efficiency of OLEDs is close to 100%, the light extraction efficiency (out-coupling efficiency) of conventional OLEDs is limited to about 20–30%. This section describes light extraction enhancement technologies for OLEDs, including achieved performance.

Generally, the destinations of emission in OLEDs are classified as external mode (air mode), substrate mode, waveguide mode (WGM), and surface plasmon mode (SPM), as shown in Fig. 9.5.

External mode means that the emitted light leaves the OLED device. In the substrate mode, the emitted light travels inside the substrate. The substrate mode is a phenomenon where total internal reflection occurs at the interface between substrate and air due to the difference in refractive indexes of the substrate and air. In waveguide mode, the emitted light travels inside the organic layers and the transparent electrode. In the surface plasmon mode excitons are quenched by the coupling into the surface plasmon associated with the metallic cathodes.

Because of these modes, in common bottom emitting OLED architectures on glass, it is estimated that only 20–30% of the total light can be extracted via the external mode, and other light disappears within the OLED device. Therefore, the light extraction enhancement (LEE) technologies are very key technologies for OLED lighting.

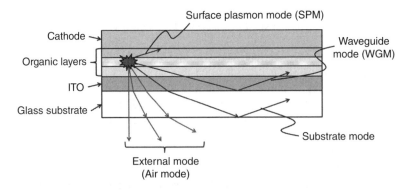

Figure 9.5   The destinations of emission in OLED devices

For extracting light from the substrate mode, the most commonly used technology is the addition of a special optical layer on the substrate. The optical layer should have such a function that the direction of the incoming light to the substrate is changed, as shown in Fig. 9.6. Such layers use, for example, microlens arrays (MLA) [6–11], micro structure films [12], or sandblasted surfaces [13].

Möller and Forrest reported that ordered microlens arrays with 10 μm diameter attached to glass substrate can increase the light extraction efficiency by a factor of 1.5 over unlensed substrates [6]. The device structure is shown in Fig. 9.7.

Galeotti et al. investigated the effect of microlens arrays for light extraction enhancement [10]. Their results are shown in Table 9.3.

Lin et al. investigated the light extraction efficiency of OLED devices attached to a commercial diffuser film (Taiwan Keiwa Inc., BS-702) or a brightness-enhancement film, BEF (3 M, VikutiᵀᴹBEF II 90/50) [12]. When attaching the films to a commercial white OLED, the luminous current efficiencies of the OLED with an attached diffuser film or BEF are reported to increase by 34% and 31%, respectively.

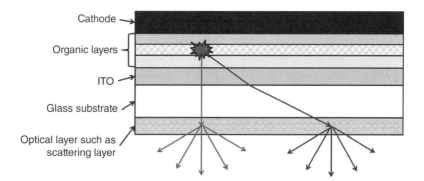

Figure 9.6   Light extraction enhancement by attaching non-uniform and/or non-flat structure to the substrates

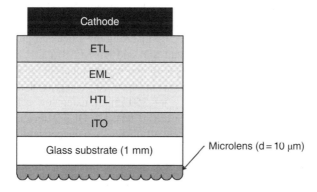

Figure 9.7   OLED device with microlens arrays [6]

Table 9.3   The effect of microlens array [10]

Diameter (µm)	SD	Height (µm)	RMS	θ (°)	Enhancement (%)
1.2	0.05	0.3	0.09	53	19
1.8	0.32	0.4	0.16	48	17
3.0	0.16	0.6	0.19	43	22
6.0	1.12	2.5	0.58	80	32

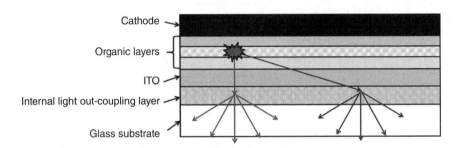

Figure 9.8   OLED device with an internal light out-coupling layer

However, these methods cannot extract the remaining light confined within thin transparent electrode and organic layers. For this issue, it was reported that some internal light out-coupling layers such as scatterings layers and diffraction gratings were fabricated at the interface between the glass substrate and the transparent electrode for reducing the waveguide modes as shown in Fig. 9.8. The internal light out-coupling layers can reduce the waveguide mode, preventing total reflection at the interface between the transparent electrode and the glass substrate by changing the direction of the light.

Examples of such technologies are the insertion of silica aerogel as a low refractive index film [14], Bragg grating [15, 16], Bragg scattering induced periodic microstructure [17, 18], and periodic photonic crystal (PC) structure [19–23].

Tsutsui et al. reported on light extraction enhancement by the insertion of a silica aerogel layer into OLED devices [14]. The device structure is shown in Fig. 9.9. The silica aerogels were prepared by the sol-gel method and showed extremely low reflective indexes of 1.01–1.10. The EQE of the device is 1.39%, while that of the reference device without the aerogel is 0.765%. The enhancement factor obtained by inserting the silica aerogel layer is 1.8.

Do et al. of Kookmin University, Samsung SDI and Korea Advanced Institute of Science and Technology investigated OLED devices with two-dimensional SiO_2/SiN_x photonic crystal (PC) layers [19]. The device structure is shown in Fig. 9.10. As a typical case, they fabricated the PC pattern with a lattice constant Λ_{cutoff} of 350nm by using low reflective index SiO_2 (n=1.48) and high reflective index SiN_x (n=1.90–1.95). The OLED device with this substrate was reported to show 1.52 times higher current efficiency than the conventional OLED device.

The light extraction enhancement technologies for surface plasmon mode have also been investigated [24–28]. For example, Hobson et al. reported that power loss to surface plasmon modes can be recovered through the use of an appropriate periodic microstructure, as shown in Fig. 9.11 [24].

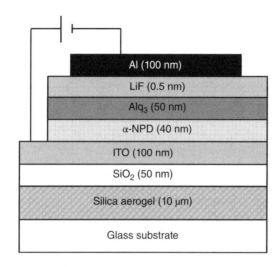

Figure 9.9   OLED with silica aerogel layer [14]

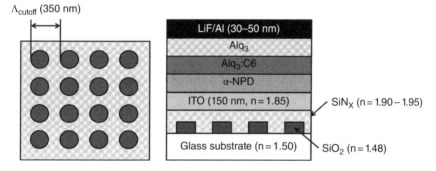

Figure 9.10   OLED devices with two-dimensional SiO2/SiNx photonic crystal (PC) layers [19]

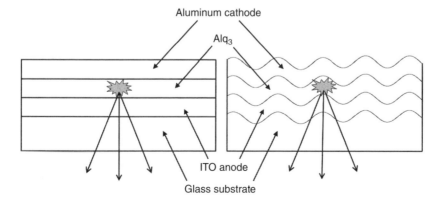

Figure 9.11   Appropriate periodic microstructure reducing surface plasmon [24]

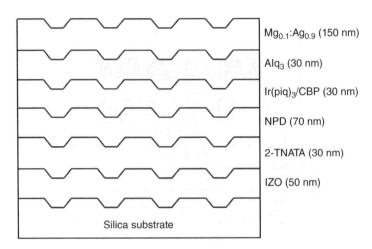

Mg$_{0.1}$:Ag$_{0.9}$ (150 nm)

Alq$_3$ (30 nm)

Ir(piq)$_3$/CBP (30 nm)

NPD (70 nm)

2-TNATA (30 nm)

IZO (50 nm)

Silica substrate

Figure 9.12   Device structure with plasmonic structure. [25]

Okamoto reported on light extraction from OLEDs with plasmonic structure [25]. The device structure in their report is shown in Fig. 9.12. He fabricated the plasmonic crystal structure with a variety of nominal depths (0–80 nm) by using a colloidal lithography technique in which homogeneous silica particles with a few hundred nanometers diameter were deposited onto a silica substrate by the Langmuir–Blodgett technique and the substrates were etched by reactive ion etching with the mask of the particles. The maximum improvement factor in terms of power efficiency compared with the flat device with the nominal depth of 0 nm was reported to be 2.35 in the device with the nominal depth of 60 nm.

As an alternative approach, Murano et al. of Novaled AG reported that a strong reduction on the plasmon absorption losses was obtained by adding a surface corrugating material to an electron transport layer [26]. The SEM image shows that the surface of the cathode layer is not flat because the surface corrugation is replicated in the cathode layer.

Shibanuma et al. of JX Nippon Oil & Energy Corporation reported light extraction technologies using corrugated substrates with a self-assembled quasi-periodic structure [27]. Using spontaneously self-assembled block copolymers (BCPs), they fabricated a quasi-periodic structure with sub-micron size. They reported achieving enhancement factors of 1.5–2.7.

Youn et al. of So's group in University of Florida (USA) reported to achieve EQE of 63.2% by extracting the surface plasmon mode with a corrugated structure and also by using a micro lens array [11].

Komoda et al. achieved high out-coupling efficiency η_{ext} of about 47%, also achieving high efficiency of 85 lm/W [29].

Another approach is a utilization of molecular orientation of organic materials in emission layers. As described in Section 4.5, Frischeisen et al. reported a significant enhancement of light out-coupling efficiency by 45%, using the effect of molecular orientation of emitting materials [30].

9.5 Performance of OLED Lighting

The performance of OLED lighting has been steadily improved. For example, LG Chem announced a commercialized OLED lighting panel, that achieved 80 lm/W, 3000 K, CRI > 80, LT_{70} = 20,000 h, in their 100 × 100 mm OLED lighting panel [31]. The efficiency is comparable with fluorescent lamps, but is still insufficient compared to LEDs.

However, the efficiency of developed white OLED devices has been actively improved, as shown in Fig. 9.13.

Table 9.4 shows several OLED lighting devices at the developmental stage. OLEDs have achieved efficiency of over 130 lm/W and lifetimes longer than 40,000 hours.

9.6 Color Tunable OLED Lighting

The color tunable lighting panel is an interesting technology from the lighting design point of view. In particular, it should be noted that the color change of lighting tends to have a significant influence on human emotion and the circadian rhythm. Color tunable lighting can be achieved by using white OLEDs with RGB stripe patterns, where each of the stripe patterns is controlled by individual current drivers. Figure 9.14 shows the schematic illustration of a color tunable OLED lighting panel with RGB side-by-side stripe patterns. This technology is a modified version of the passive-matrix display.

Pioneer and Mitsubishi Chemical have developed color tunable OLED lighting panels with RGB stripe patterns [38, 39]. Ohshita of Pioneer reported on the technologies in Japanese literature [40]. They achieved the power efficiency of 50 lm/W (at 1000 cd/m²) and the lifetime T_{70} of 8000 hours (at the initial luminance of 2000 cd/m²), combining an out-coupling film.

Weaver et al. of Universal Display Corporation (UDC) and Acuity Brands Lighting developed a color tunable phosphorescent white OLED lighting panel [41]. The performance of their technologies is summarized in Table 9.5.

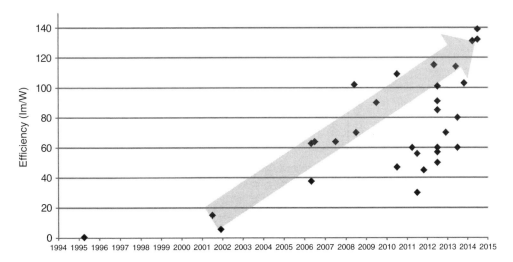

Figure 9.13 Progress of efficiency of white OLED devices

Table 9.4 Several OLED lightings in the developmental level

Affiliation	Year	Luminance (cd/m²)	Efficiency (lm/W)	CRI	CIE	Technologies	Comments	Ref.
UDC	2008	1	102			LEE		[32]
Leo	2009		90			LEE		[33]
UDC	2010	1000	109	80	(0.428, 0.421)	LEE	3295 K, $LT_{70} = 15{,}000$ h	[34]
Konica Minolta	2014		139				$T_{50} = 55{,}000$ h (@1000 cd/m²)	[35]
Panasonic	2014		133	84	(0.48, 0.43)	LEE	10×10 cm	[36]
First-Lite	2014		111.7	85		LEE, MPE(3)		[37]

CRI = color rendering index.

LEE = light extraction enhancement technology.

MPE = multi-photon emission. The number in the parentheses is the number of stacking units.

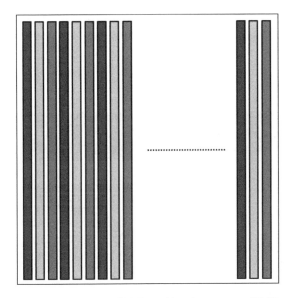

Figure 9.14 Color tunable OLED lighting with strip patterns of RGB emission [40]

Table 9.5 Example of color tunable phosphorescent OLED lighting panel [41]

	Blue	Green	Red
1931 CIE (x,y)	(0.17, 0.39)	(0.41, 0.57)	(0.64, 0.36)
Current efficiency (cd/A)	53	74	30
Voltage (V)	4.7	2.6	3.0
Power efficiency (lm/W) at $1000\,cd/m^2$	35	89	32

9.7 Application of OLED Lighting – Products and Prototypes

The OLED lighting business must compete with other forms of lighting in the market. Since OLED lighting possesses high potential for the future of lighting due to such unique features as planar emission, thinness, light weight, high efficiency, possibility of flexibility, high color rendering index, the business growth of OLED lighting is predicted. Examples of OLED lighting panels are shown in Fig. 9.15 [42].

Fuji Chimera Research Institute, Inc. forecasts that the OLED lighting market will grow to 500 million USD in 2020, while the market in 2011 was only 2.5 million USD [43].

IDTechEX predicted that the OLED lighting market growth will be very slow until 2019, where the sales at panel level will remain below 200 million USD globally. However, they also forecast that the market would grow to 1.9 billion USD in 2025 as an optimistic scenario [44].

The global commercial lighting market is about 40 billion USD in 2013 and is predicted to be grow to about 50 billion USD in 2020 [45, 46]. Compared to such a huge market, the predicted market for OLED lighting in 2020 is still tiny. The predicted OLED lighting market of 1.9 billion USD in 2025 is not small but is only few percent of the total lighting market. Therefore, this suggests that a big market awaits OLED lighting.

Since the current cost of OLED lighting is too high to compete with LED lighting in cost-competitive mass commercial products, OLED lighting needs to penetrate niche markets, based on its unique features.

For example, Takahata Electronics Corporation, Yamagata, Japan, has developed and commercialized OLED lighting products for medical uses [47]. Some products are shown in Figs 9.16 and 9.17. One of the unique products is a nurse-light with color tunable OLED lighting, shown in Fig. 9.16. At night in hospitals, nurses presently use a pen-light. However,

Figure 9.15   Some OLED lighting products [42]. (provided by Lumiotec Inc.)

Figure 9.16   Nurse light [47]. (provided by Takahata Electronics Corporation)

Figure 9.17 An example of OLED lighting products [47]. (provided by Takahata Electronics Corporation)

Figure 9.18 Cosmetic OLED lighting. (provided by Pioneer Corporation)

Figure 9.19 Examples of products and prototypes of OLED lightings. (provided by Kaneka Corporation)

the problem of the pen light is disturbing the patients owing to glare from its strong point light source. On the other hand, the nurse-light using OLED lighting is gentle for the patient, not disturbing or waking them up. Also, it is possible to check the color of the patient's face and skin by using the nurse-light, because the OLED lighting is a planar light. Moreover, by using the nurse-light, both hands of nurse are free, helping their work.

OLED lighting has started to be utilized in museum because it causes no damage to artifacts due to the absence of UV light.

Some other products and prototypes of OLED lighting devices are shown in Figs 9.18 and 9.19, the former showing cosmetic lighting.

In the business competition with other forms of lighting such as LED lighting, the ability to be flexible seems to be very important, because flexibility can realize unique designs and unique applications. Flexible OLED lighting technologies will be described in Section 10.4.

References

[1] News release of Lumiotec, 24 July 2011. www.lumiotec.com/pdf/110727_LumiotecNewsRelease%20JPN.pdf
[2] J. Kido, K. Hongawa, K. Okuyama and K. Nagai, *Appl. Phys. Lett.*, **64(7)**, 815–817 (1994).
[3] J. Kido, M. Kimura and K. Nagai, *Science*, **267**, 1332–1334 (1995).
[4] K. Ghaffarzadeh, N. Bardsley, "OLED *Lighting Opportunities 2014–2025: Forecasts, Technologies, Players*", IDTechEX (2014).
[5] United States Department of Energy, "Solid-State Lighting Research and Development" Multi-Year Program Plan, April 2014.
[6] S. Möller, S. R. Forrest, *J. Appl Phys.*, **91(5)**, 3324–3327 (2002).
[7] S.-H. Eom, E. Wrzesniewski, J. Xue, *Org. Electron.*, **12**, 472–476 (2011).
[8] M. K. Wei, H. Y. Lin, J. H. Lee, K. Y. Chen, Y. H. Ho, C. C. Lin, C. F. Wu, H. Y. Lin, J. H. Tsai, T. C. Wu, *Opt. Commun.*, **281**, 5625–5632 (2008).
[9] M. K. Wei, I. L. Su, *Opt. Express.*, **12**, 5777–5782 (2004).
[10] F. Galeotti, W. Mróz, G. Scavia, C. Botta, *Organic Electronics*, **14**, 212–218 (2013).
[11] W. Youn, J. Lee, M. Xu, C. Xiang, R. Singh, F. So, *SID 2014 Digest*, 5.3 (p. 40) (2014).
[12] H.-Y. Lin, J.-H. Lee, M.-K. Wei, C.-L. Dai, C.-F. Wu, Y.-H. Ho, H.-Y. Lin, T.-C. Wu, *Optics Communications*, **275**, 464–469 (2007).
[13] J. Zhou, N. Ai, L. Wang, H. Zheng, C. Luo, Z. Jiang, S. Yu, Y. Cao, J. Wang, *Org. Electron.*, **12**, 648–653 (2011).
[14] T. Tsutsui, M. Yahiro, H. Yokogawa, K. Kawano, M. Yokoyama, *Adv. Mater.*, **13(15)**, 1149–1152 (2001).
[15] J. M. Lupton, B. J. Matterson, D. Ifor, W. Samuel, M. J. Jory, W. L. Barnes, *Appl. Phys. Lett.*, **77**, 3340 (2000).
[16] J. M. Ziebarth, A. K. Saafir, S. Fan, M. D. McGehee, *Adv. Funct. Mater.*, **14**, 451–456 (2004).
[17] J. M. Lupton, B. J. Matterson, I. D. W. Samuel, M. J. Joy, W. L. Barnes, *Appl. Phys. Lett.*, **77**, 3340 (2000).
[18] A. Kock, E. Gornik, M. Hauser, K. Beinstingl, *Appl. Phys. Lett.*, **57**, 2327 (1990).
[19] Y. R. Do, Y-C. Kim, Y-W. Song, Y.-H. Lee, J. Appl. *Phys.*, **96(12)**, 7629–7636 (2004).
[20] A. O. Altun, S. Jeon, J. Shim, J. H. Jeong, D. G. Choi, K. D. Kim, J. H. Choi, S. W. Lee, E. S. Lee, H. D. Park, J.R. Youn, J. J. Kim, Y. H. Lee, J. W. Kang, *Org. Electron.*, **11**, 711–716 (2010).
[21] K. Ishihara, M. Fujita, I. Matsubara, T. Asano, S. Noda, H. Ohata, A. Hirasawa, H. Nakada, N. Shimoji, *Appl. Phys. Lett.*, **90**, 111114 (2007).
[22] S. Jeon, J. W. Kang, H. D. Park, J. J. Kim, J. R. Youn, J. H. Jeong, D. G. Choi, K. D. Kim, A. O. Altun, Y. H. Lee, *Appl. Phys. Lett.*, **92**, 223307 (2008).
[23] M. Fujita, K. Ishihara, T. Ueno, T. Asano, S. Noda, H. Ohata, T. Tsuji, H. Nadaka, N. Shimoji, *Jpn J. Appl. Phys.*, **44(6A)**, 3669–3677 (2005).
[24] P. A. Hobson, S. Wedge, J. A. E. Wasey, I. Sage, W. L. Barnes, *Adv. Matter.*, **14(19)**, 1393–1396 (2002).
[25] T. Okamoto, *Proc IDW/AD'12*, OLED1-3 (2012); T. Okamoto, K. Shinotsuka, *Appl. Phys. Lett*, **104(9)**, 093301 (2014).
[26] S. Murano, D. Pavicic, M. Furno, C. Rothe, T. W. Canzler, A. Haldi, F. Löser, O. Fadhel, F. Cardinali, O. Langguth, *SID 2012 Digest*, 51.2 (p. 687) (2012).

[27] T. Shibanuma, T. Seki, S. Toriyama, S. Nishimura, *SID 2014 Digest*, P-152 (p. 1554) (2014).

[28] J. Frischeisen, Q. Niu, A. Abdellah, J. B. Kinzel, R. Gehlhaar, G. Scarpa, C. Adachi, P. Lugli, W. Brütting, *Optics Express*, **19**, A7–A19 (2011).

[29] K. Yamae, H. Tsuji, V. Kittichungchit, Y. Matsuhisa, S. Hayashi, N. Ide, T. Komoda, *SID 2012 Digest*, 51.4 (p. 694) (2012).

[30] J. Frischeisen, D. Yokoyama, A. Endo, C. Adachi, W. Brütting, *Organic Electronics*, **12**, 809–817 (2011).

[31] LG Chem, Press release, 30 September 2013: www.lgchem.com/global/lg-chem-company/information-center/press-release/news-detail-567

[32] B. W. D'Andrade, J. Esler, C. Lin, V. Adamovich, S. Xia, M. S. Weaver, R. Kwong, J. J. Brown, *Proc. IDW'08*, OLED1-4L (p. 143) (2008).

[33] S. Reineke, F. Lindner, G. Schwartz, N. Seidler, K. Walzer, B. Lu¨ssem1, K. Leo, *Nature*, **459**, 234–238 (2009).

[34] P. A. Levermore, V. Adamovich, K. Rajan, W. Yeager, C. Lin, S. Xia, G. S. Kottas, M. S. Weaver, R. Kwong, R. Ma, M. Hack, J. J. Brown, *SID 10 Digest*, 52.4 (p. 786) (2010).

[35] Oral presentation by T. Tsujimura at the session 10.1 of SID 2014 (2014); T. Tsujimura, J. Fukawa, K. Endoh, Y. Suzuki, K. Hirabayashi, T. Mori, *SID 2014 Digest*, 10.1 (p. 104) (2014).

[36] K. Yamae, V. Kittichungchit, N. Ide, M. Ota, T. Komoda, *SID 2014 Digest*, 47.4 (p. 682) (2014).

[37] Y.-S. Tyan, Y.-X. Shen, J.-J. Peng, L. Z., C. Feng, Y.-W. Sui, H. Lu, Y.-C. Wu, H.-J. Ren, Q.-H. Tian, X.-Y. Gu, G.-Y. Huang, *SID 2014 Digest*, 47.2 (p. 675) (2014).

[38] News release of Pioneer Corporation, 3 Jun 2013 (in Japanese). http://pioneer.jp/corp/news/press/index/1636.

[39] News release of Mitsubishi Chemical Corporation, 3 Jun 2013 (in Japanese). www.m-kagaku.co.jp/newsreleases/2013/20130603–1.html.

[40] I. Ohshita, *Pioneer T&D*, **22**, 24–32 (2013).

[41] M. S. Weaver, X. Xu. H. Pang, R. Ma, J. J. Brown, M.-H. Lu, *SID 2014 Digest*, 47.1 (p. 672) (2014).

[42] see homepage of Lumiotec Inc. www.lumiotec.eu/index-en.html.

[43] Fuji Chimera Research Institute, Inc., *"Report on the future's possibility of flexible/transparent/printed electronics"* (2012).

[44] K. Ghaffarzadeh, N. Bardsley, "OLED *Lighting Opportunities 2014–2025: Forecasts, Technologies, Players"*, IDTechEX (2014).

[45] *"Energy Efficient Lighting for Commercial Markets"*, prepared by Navigant Research, 2Q 2013.

[46] United States Department of Energy, *Solid-State Lighting Research and Development, Multi-Year Program Plan*, April 2014.

[47] see homepage of Takahata Electronics Corporation. www.takahata-denshi.co.jp/.

10

Flexible OLEDs

Summary

One of the significant advantages of OLEDs is that they can be fabricated on flexible substrates. The use of flexible substrates instead of conventional glass substrates can significantly reduce the thickness and weight of displays and lighting. In addition, flexible OLEDs bring about additional attractive features from the product design point of view, because flexible OLEDs can provide such unique designs as curved, bent, folded, rolled, and ultimately flexible. Moreover, use of flexible substrates has great potential of production innovation such as utilization of roll-to-roll (R2R) process with low mass-production cost. As the candidates for flexible substrates, ultra-thin glasses, stainless steel foils and plastic films are well known. This chapter describes current status and future potential of three types of flexible substrates and their applications to flexible OLED displays and lightings.

Key words

flexible, ultra-thin glass, stainless steel foil, plastic film, barrier, planarization, flexible display, flexible lighting

10.1 Early Studies of Flexible OLEDs

Flexible OLEDs have been studied and reported since the early stages of OLED research and development.

In 1992, Gustafsson et al. of Uniax Corporation (USA) reported a flexible OLED that consists of a polyethylene terephthalate (PET) film (thickness 100 μm) as substrate, a thin film of polyaniline (PANI) hole-injecting electrode (200 Ω/sq, transmittance of 70% at visible wavelength), a film of poly(2-methoxy, 5-(2′-ethyl-hexoxy)-1,4-phenylene-vinylene)

OLED Displays and Lighting, First Edition. Mitsuhiro Koden.
© 2017 John Wiley & Sons, Ltd. Published 2017 by John Wiley & Sons, Ltd.

(MEH-PPV) as the light emitting polymer and calcium as the electron-injecting cathode [1]. They reported that the device showed the EQE of 1%.

In 1997, Gu et al. of Princeton University and University of South California (USA) fabricated a flexible OLED by vacuum deposition on a transparent thin plastic film with an ITO electrode [2]. They reported that the flexible OLED performance is comparable with that of conventional OLEDs deposited on glass substrates and does not deteriorate after repeated bending.

In 1997, Wu et al. of Forrest's group at Princeton University (USA) reported an amorphous-Si-TFT/OLED display with the size of 4×4 cm [3]. They used grade 430 stainless steel foils with thicknesses of 76–230 μm. The structure of the OLED part was Pt/PVK:PBD:C6/Mg:Ag/ITO, where PVK is hole-transport matrix polymer poly(N-vinylcarbazole), PBD is host electron-transport material 2(-4-biphenyl)-5-(4-tert-butyl-phenyl)-1,3,4-oxadiazole, and C6 is coumarin 6. The display was reported to show a brightness of 100 cd/m^2 and could withstand considerable mechanical stress, such as being dropped 30 feet onto concrete and bending.

10.2 Flexible Substrates

Flexible substrates are one of the most important technologies for realizing flexible OLEDs. Among the properties of flexible substrates, gas barrier is significant. In the discussion of the gas barrier property, one of the most often used parameters is water vapor transmission rate (WVTR), which is represented by g/m^2/day. Figure 10.1 shows the required WVTR of flexible substrates for various applications. The required WVTR is relatively high in such applications as packaging for foods, medicines, and electronics components. On the other hand, electronics devices such as LCD, LED, PV, OPV, and OLED require low WVTR. In particular, OLEDs require very low WVTR because organic layers and/or electrodes are easily damaged by water. It is often said that WVTR of the order of 10^{-6} is required for flexible OLED devices.

This section describes and compares three types of flexible substrates (ultra-thin glass, stainless steel foil, and plastic film). Table 10.1 shows the comparison of the three types of flexible substrates. Each substrate has its own attractive features and intrinsic issues.

Glass is the most commonly used substrate in OLED displays and OLED lighting. The commonly used glass substrates with a thickness of 0.5 mm or 0.7 mm do not have flexibility.

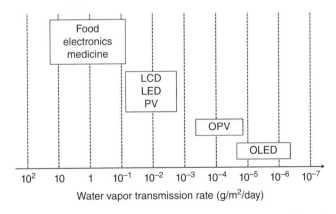

Figure 10.1   Required water vapor transmission rate of flexible substrates for various applications

Table 10.1 The properties of three types of flexible substrates

	Ultra-thin glass	Stainless steel foil	Plastic film
Specific gravity	~2.5	~7.8	~1.4
Temperature resistance	Excellent	Excellent	Low
	Tg: 600 °C	Tg: 1400 °C	PET: 110 °C (Tg)
			PEN: 180 °C (Tg)
CTE	Excellent	Good	Poor
	3–8 ppm/°C	14–16 ppm/°C	PEN: 18–20 ppm/°C
			COP: 60–65 ppm/°C
Surface smoothness	Excellent	Not smooth	Not smooth
Problems in handling	Breakable	Conductivity	Less rigid
Roll-to-roll	Poor experience	Poor experience	Much experience
Vapor barrier	Excellent	Excellent	Poor

CTE = coefficient of thermal expansion, COP = cyclic olefin polymer

However, by reducing the thickness, glass substrates turn into flexible substrates. Indeed, ultra-thin glass with a thickness of 100 μm or 50 μm clearly shows flexibility. Such ultra-thin glass substrates have several advantages: excellent temperature resistance, excellent chemical resistances, low coefficient of thermal expansion, excellent surface smoothness, and excellent vapor barrier properties, which are intrinsic advantages of glasses. However, since glass is fragile, handling techniques in production using ultra-thin glass are still problematic. Therefore, experience of the application of ultra-thin glass to R2R equipment has not been very widespread.

Stainless steel foils also become flexible substrates when the thickness is reduced. Stainless steel foil has several advantages such as excellent temperature resistance and excellent vapor barrier properties, which are intrinsic advantages of stainless steels. However, the surface of bare stainless steel foils is not very smooth. In addition, since stainless steel is a conductive material, insulating technologies are required for applying OLED devices.

From a general R2R production point of view, plastic films have been widely used, but when used for OLEDs, plastic films have several problems, the most serious of which is poor gas barrier property. In addition, temperature resistance, chemical resistance, surface smoothness, coefficient of thermal expansion, etc. of plastic films are still issues.

One of the interesting alternative approaches for flexible films is cellulose nanofiber (CNF) film [4, 5], which is a kind of paper, but it is transparent. While normal papers are not transparent due to the micrometer order of the cellulose fibers, CNF can be transparent due to the nanofibers being as small as 15 nm. In addition, CNF has several advantages such as low coefficient of thermal expansion (e.g. <10 ppm/°C), high Young's modulus (e.g. >11 GPa), high temperature resistance (e.g. no glass transition point in the range of 0–250 °C).

10.2.1 Ultra-Thin Glass

It is usually said that substrates for OLED devices are required to have a high gas barrier property such as 10^{-6} g/m^2/day, as shown in Fig. 10.1. While it is not easy for plastic films to achieve such a high gas barrier property, glass can easily achieve the value even if it is as thin as 50 μm.

This ultra-thin glass can be made by an overflow method and can be applied to a roll. For example, Nippon Electric Glass has developed a roll of ultra-thin glass with thickness of 50 μm. [6]. An example is shown in Fig. 10.2.

Figure 10.3 shows the relationship between the curvature radius of bending R and the stress at the top of the curvature, σ [6]. The stress of the top of the curvature, σ, is said to be related to the curvature radius of bending R by the following equation.

$$\sigma = ET/2R$$

where E is Young's modulus and T is the glass thickness. This equation and Fig. 10.3 clearly indicate that flexibility increases with decreasing the glass thickness. It is said that glass is

 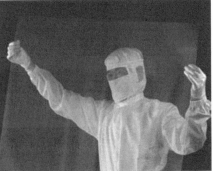

Figure 10.2   An ultra-thin glass roll [6]. (provided by Nippon Electric Glass Co. Ltd)

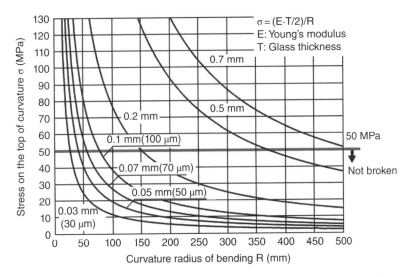

Figure 10.3   The relationship between the curvature radius R of bending and the stress of the top of the curvature σ [6]. (Source: Nippon Electric Glass Co. Ltd)

broken by a stress higher than 50 MPa, although breaking stress is also affected by the edge condition of the glass. Therefore, 100 μm ultra-thin glass can withstand the a radius larger than about 70 mm and a 50 μm ultra-thin glass can withstand a radius down to about 40 mm.

On the other hand, one of the serious problems of ultra-thin glass is handling difficulty due to its fragility.

Asahi Glass (Japan) reported the carrier glass lamination technology [7]. In this technology, for example, 0.5 mm thick carrier glass is laminated to 0.1 mm thick ultra-thin glass by using a media layer, as shown in Fig. 10.4. This laminated substrate has good thermal, chemical, and mechanical stabilities and can be applied to conventional fabrication equipment for OLEDs. After fabricating and assembling the devices, the carrier glass is separated from the thin flexible device.

Cutting of glass is also closely related to the fragile feature of glass. If the cutting edge has a defect, the glass is easily broken in the device fabrication process. Nippon Electric Glass reported a new cutting technology to solve this problem [8]. The technology is laser fusing cutting, using a CO_2 laser, giving a fused glass edge on the ultra-thin glass.

Kobe Steel (Japan) developed a roll-to-roll DC magnetron sputtering deposition system to which ultra-thin glass can be applied [9]. In their experiments, a 10 wt% SnO_2-doped ITO target with a width of 120 mm and length 500 mm was used, and a 30 kHz-pulsed DC power was supplied for stable magnetron discharge. The temperature of the main drum was set at 300 °C. They used a roll-to-roll ultra-thin glass with a thickness of 50 μm, width 300 mm, and length 10 m. They reported achieving a sheet resistance of 7.5 Ω/sq, and a transmittance of 83.2% in the deposited ITO film with the thickness of 190 nm.

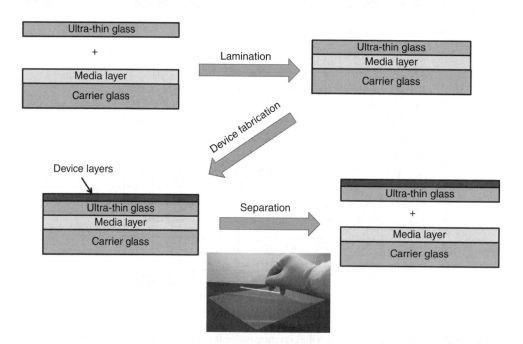

Figure 10.4 Ultra-thin glass with carrier glass [7]. (Source: Asahi Glass Co., Ltd)

10.2.2 Stainless Steel Foil

Stainless steel foil has been recognized as a candidate for an alternative substrate to glass, because it has several attractive properties such as excellent gas barrier, excellent temperature resistance, excellent mechanical strength, and light weight.

However, stainless steel foil has such problems as surface roughness and a requirement for insulation. The poor surface flatness easily gives rise to electrical leakage between anode and cathode in OLED devices.

In order to solve these problems of surface smoothness and insulation, film coating technologies on a stainless steel foil have been investigated. The coating film should have high heat resistance and gas barrier property but without spoiling the advantages of stainless steel foil.

Xie et al. of City University Hong Kong (China) and Institute of Materials Research and Engineering (Singapore) applied a spin-on-glass (SOG) film on stainless steel foil with the thickness of 20 μm [10]. They reported that the 1 μm thick SOG film served as a planarization/ insulating layer on the conductive and rough surface of steel. They also reported that the SOG film showed superior film thickness uniformity, strong adhesion to the metal surface and process capability at elevated temperature. They fabricated a top emitting OLED device on this stainless steel foil. The device structure and the emission picture of the top emitting OLED device on stainless steel foil are shown in Fig. 10.5. The device was reported to show 4.4 cd/A, which is better than the 3.7 cd/A obtained in the bottom emitting OLED with the same organic layers.

Yamada et al. of Nippon Steel and Sumitomo Metal Corporation Group reported applying inorganic–organic hybrid materials prepared by the sol-gel method on stainless steel foils [11]. They coated the material on a stainless steel foil 50 μm thick. While the bare stainless steel foil has a very rough surface with Ra = 10.8 nm, the coated one has a smooth surface with Ra = 2.1 nm, as shown in Fig. 10.6. Stainless steel foil is useful for not only substrates of OLED devices but also encapsulating substrates for OLEDs because it has excellent mechanical stability and gas barrier properties. In the Yamagata University Organic Thin Film Device Consortium, OLED lighting devices fabricated on an ultra-thin glass were encapsulated by a stainless steel foil [12]. This is described in Section 10.4.1.

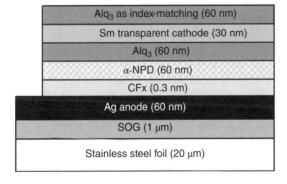 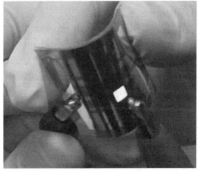

Figure 10.5 Device structure and an emission picture of a top emitting OLED device on stainless steel foil [10]

Figure 10.6 Bare and coated stainless steel foils from Nippon Steel and Sumitomo Metal Corporation Group

10.2.3 Plastic Films

Candidates for plastic films for OLED devices are such as PET, PEN, and PI. The molecular structures of these films are shown in Fig. 10.7. PET (polyethylene terephthalate) is an inexpensive film. However, the problem with PET is low temperature tolerance, because the glass transition temperature (Tg) is around 80 °C. PEN (polyethylene naphthalate) has a higher temperature tolerance (Tg ~ 155 °C) than PET but it still requires a low process temperature for the TFT process, and it is more expensive than PET. PI (polyimide) films have tolerance to TFT process because the Tg can be higher than 300 °C. However, one of the serious problems of polyimide films is coloration.

In plastic films, vapor barrier technologies are very important because plastic films do not have good enough barrier properties for OLED devices.

While the required barrier properties for various applications are shown in Fig. 10.1, the WVTR values of polymers are several to several tens of $g/cm^2/day$. This means that the required barrier properties for some vapor-sensitive foods can be achieved, if low or medium barrier technologies are applied to films. However, it is commonly said that OLEDs require very high barrier properties such as $10^{-6}\,g/cm^2/day$. In order to achieve such high barrier properties, various technologies have been investigated and developed.

Roughly speaking, there are two approaches in order to improve the barrier property of plastic films. One is high quality, low defect inorganic single layer and the other is multi-layer stacks in which inorganic barrier layers and organic (polymer) interlayers are usually deposited alternately.

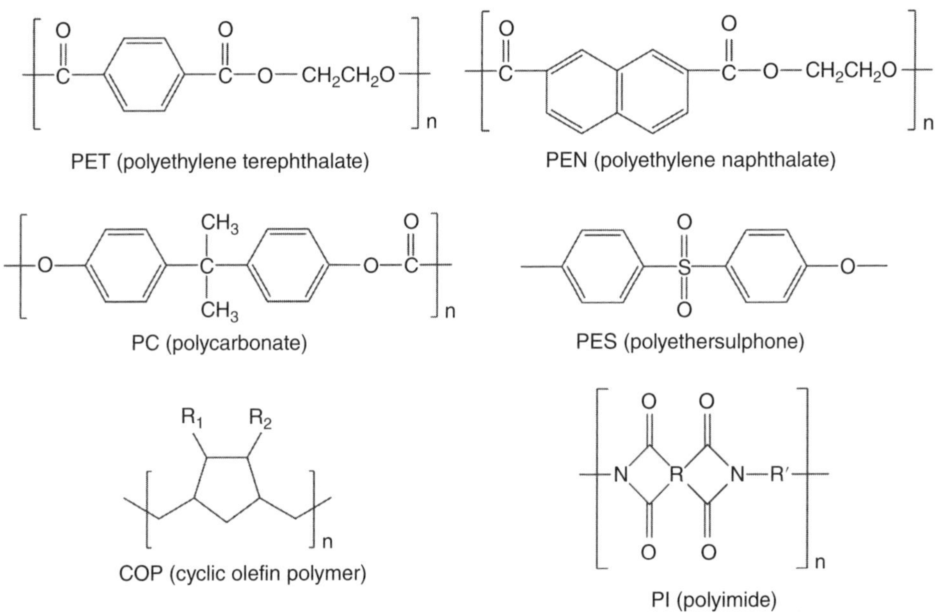

PET (polyethylene terephthalate)

PEN (polyethylene naphthalate)

PC (polycarbonate)

PES (polyethersulphone)

COP (cyclic olefin polymer)

PI (polyimide)

Figure 10.7   Molecular structures of some plastic films for flexible OLED devices

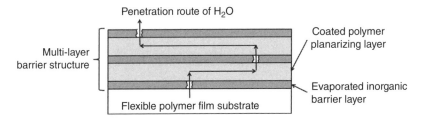

Penetration route of H_2O

Multi-layer barrier structure

Coated polymer planarizing layer

Flexible polymer film substrate

Evaporated inorganic barrier layer

Figure 10.8   Concept of multi-layer barrier structure

As barrier layers, inorganic layers such as silicon oxide, silicon nitride, and aluminum oxide are often deposited on plastic films by sputtering, chemical vapor deposition (CVD), atomic layer deposition (ALD), etc.

However, it is not so easy to prevent moisture penetration by using the monolayer, because these layers tend to have defects such as pinholes, cracks, and contaminated particles. Therefore, multi-layers have been widely investigated because multi-layer barriers have such advantages as coverage of defects, lengthened path of moisture diffusion (so-called "tortuous path model") and improved mechanical stability. The schematic concept of the multi-layer barrier structure is shown in Fig. 10.8.

As a barrier technology, Burrows et al. of Pacific Northwest National Laboratory (USA) and Vitex Systems Inc. (USA) have developed a multi-layer structure consisting of vacuum deposited polyacrylate layers between multiple layers of vapor-barrier material such as Al_2O_3 [13]. They reported that the archived barrier properties were $<0.005\,cm^3/m^2/day$ for O_2 and $<0.005\,g/m^2/day$ for H_2O at $38\,°C$.

Weaver et al. of Universal Display Corporation and Pacific Northwest National Laboratory (USA) developed a 175 μm thick PET film with an organic–inorganic multi-layered barrier film, achieving an estimated WVTR of 2×10^{-6} g/m^2/day [14]. Their multi-layer is alternating layers of polyacrylate films and a 10–30 nm thick Al$_2$O$_3$.

Recently, Suzuki et al of Lintec Corporation (Japan) developed multi-layer barrier films fabricated by wet coating technologies of barrier-precursor and plasma assisted surface modifications [15]. The barrier layers are fabricated by wet coating of ceramic precursor and plasma surface modification and a second barrier layer is fabricated on the first one. It is expect that wet coating of the second barrier precursor could cover defects in the first barrier layer. In a three-layer structure, they obtained 10^{-5} g/m^2/day under 40 °C, 90%RH.

Tsujimura et al. of Konica Minolta (Japan) developed two barrier films that showed WVTRs of 5.9×10^{-5} and 6.9×10^{-6} g/m^2/day, respectively [16]. When OLED devices were fabricated on these barrier films, very little or almost no dark spot growth was observed in 85 °C/85%RH storage condition. They announced that flexible OLED panels were being manufactured with the world's first roll-to-roll equipment for OLEDs using plastic barrier film.

As a future technology, atomic layer deposition (ALD) has attracted attention because it can deposit high density films.

Groner et al. of University of Colorado (USA) reported thin films of Al$_2$O$_3$ grown by atomic layer deposition (ALD) [17]. They reported that Al$_2$O$_3$ ALD films with thicknesses of ≥5 nm, oxygen transmission rates were below the MOCON instrument test limit of similar to 5×10^{-3} g/m^2/day. Applying a more sensitive radioactive tracer method, H$_2$O-vapor transmission rates of around 1×10^{-3} g/m^2/day were measured for single-sided Al$_2$O$_3$ ALD films with thicknesses of 26 nm on the polymers.

One of the problems of ALD is the slow deposition rate. Further development for practical uses of ALD is expected.

10.3 Flexible OLED Displays

One of the impressive early challengers for flexible OLED displays using a plastic film was a 3″ passive-matrix OLED (PM-OLED) developed by Pioneer [18, 19]. The specifications are summarized in Table 10.2.

Table 10.2 A 3″ flexible passive-matrix OLED display developed by Pioneer [18, 19]

Display size	3″ diagonal
Pixel number	160 × 120
Luminance	70 cd/m^2
Color	Full color (256 gray scale)
Device structure	Bottom emission
Driving method	Passive-matrix
Thickness	0.2 mm
Weight	3 g (including IC)

In order to fabricate the flexible display, they developed high reliability barrier layers using SiON, investigating the degradation phenomena induced by a lack of barrier properties. The developed device structure and an example of pictures are shown in Figs 10.9 and 10.10, respectively [18, 19].

The OLED device is sandwiched by two barrier layers. The one is a barrier layer on the plastic substrate, and the other is a passivation layer on the OLED device. In addition, in order to prevent the occurrence of dark spots, a UV-cure type planarization layer with the thickness of about 5 μm was coated on the plastic film because the rugged surface induces defects such as pinholes in the barrier layer.

The barrier layer on the plastic film was a SiON film fabricated by sputtering. The thickness was 200 nm. They investigated the effect of the molar ratio of O and (O+N) in the SiON film on the transmittance of the film and the barrier property, which was evaluated for the

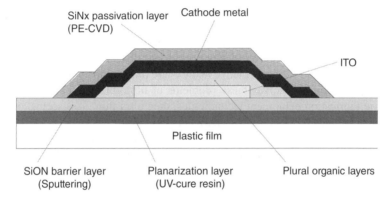

Figure 10.9 The device structure of a flexible OLED developed by Pioneer [18, 19]

Figure 10.10 A picture of a 3″ flexible PM-OLED display developed by Pioneer [18, 19]. (provided by Pioneer Corporation)

elongated non-emission area of the OLED device with the barrier SiON film. They found that the optimized ratio with a compatibility of transmittance and barrier property was the ratio O/(O+N) of 40–80%.

A SiNx passivation layer was deposited on the cathode of the flexible OLED device by plasma enhanced chemical vapor deposition (CVD). The deposition temperature needs to be low so as not to damage the OLED. The advantages of CVD are mechanical stress control and good step coverage. By optimizing process conditions of the CVD, they achieved long-term storage reliability.

Using the technologies mentioned above, they reported that no degradation was observed under 60 °C/95%RH storage test in 500 hours.

After and in parallel to their report, a lot of research and development on flexible OLED displays has taken place. This section describes them.

10.3.1 Flexible OLED Displays on Ultra-Thin Glass

As described in Section 10.2, ultra-thin glass has several advantages such as high gas barrier, excellent chemical and temperature resistance, excellent surface smoothness, and low coefficient of thermal expansion. These attractive features are fully compatible with TFT fabrication processes. Therefore, if one solves these problems of ultra-thin glass such as fragility and constructs the fabrication equipment technologies, then ultra-thin glass seems to be a strong candidate for substrates for flexible AM-OLED displays.

Kuo et al. of Chunghwa Picture Tubes (Taiwan) developed a 6″ full-color AM-OLED display with 480×640 dots on ultra-thin glass, using a-Si-TFT circuit [20]. The thickness of the ultra-thin glass was 100 μm and the thickness of the AM-OLED panel was 250 μm.

10.3.2 Flexible OLED Displays on Stainless Steel Foil

Various attempts at making OLED display fabrications on stainless steel foils have been tried. Since stainless steel foil is not transparent, displays on stainless steel foil need to use a top emitting device structure.

In 1996, Theiss and Wagner of Princeton University (USA) fabricated a-Si:H thin-film transistor display on stainless steel foils, aiming to develop backplanes for OLED displays [21].

In 1997, Wu et al. of Forrest's group at Princeton University (USA) reported on an OLED display of 4 cm×4 cm, which was driven by amorphous-Si-TFTs [3].

Troccoli et al. of Hatalis's group at Lehigh University (USA) developed poly-silicon TFT circuits on flexible stainless steel foils, achieving an electron mobility of more than 300 cm²/Vs [22].

Jin et al. of Samsung SDI reported a 5.6″ flexible full-color top emission AM-OLED display fabricated on stainless steel foil [23]. It had 160×350 pixels and the resolution was 66 ppi. Since the surface of the stainless steel foil is so rough (rms = 81.4 nm), they developed a low-cost super mirror planarization technique, obtaining a smooth surface with the rms 3.3 nm, comparable to that obtained by the expensive chemical mechanical planarization (CMP) technique. After depositing SiO_2 with a thickness of 1 μm on the stainless steel foil to insulate the poly-Si active layer from the metal substrate, the TFT array was fabricated using a conventional LTPS process. They fabricated p-channel low temperature poly-Si (LTPS)

TFTs with a field-effect mobility of $71.2\,cm^2/Vs$ on stainless steel foils. On the TFT structures, they fabricated top emission OLEDs, depositing a reflective anode, organic layers, and a transparent cathode with this sequence. As an encapsulation, a multi-layer thin-film structure provided by Vitex System Inc. was applied to the OLED structure.

Templier et al. at CEA-LETI (France) fabricated a 2.8″ diagonal flexible backplane for LTPS-OLED displays using a stainless steel foil [24]. The specifications of the backplane are pixel number of $120 \times 160\,(\times 3)$, resolution of 70 ppi, field-effect mobility of $83\,cm^2/Vs$, and current ratio of more than 10^7.

Ma et al. of Universal Display Corporation (USA) and Kyung Hee University (Korea) developed a 3″ diagonal AM-OLED display fabricated on a stainless steel foil [25]. They fabricated an a-Si TFT circuit with 2-TFT structure and combined it with a top emitting phosphorescent OLED. The specifications of the prototype display included it being $150\,cd/m^2$ and 67 ppi.

10.3.3 Flexible OLED Displays on Plastic Film

In active-matrix OLED (AM-OLED) displays fabricated on a flexible plastic film, TFT circuits need to be fabricated on a plastic film.

One typical fabrication method for flexible AM-OLED displays on a flexible plastic film is the coating/de-bonding method, schematically illustrated in Fig. 10.11. In this method, a plastic film is coated on a glass substrate. For plastic films, polyimide films are often used because they have several advantages such as high thermal stability, high glass transition temperature (Tg), and good chemical resistance. These advantages give compatibility with the TFT fabrication process. The glass substrate works as a carrier substrate in the fabrication

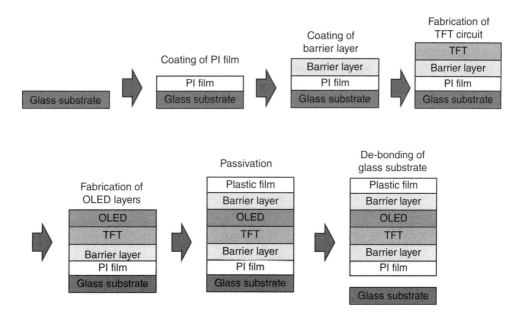

Figure 10.11 Schematic illustration of the coating/de-bonding method

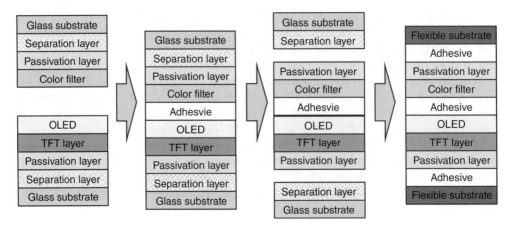

Figure 10.12   Schematic illustration of the transfer method reported by Semiconductor Energy laboratory (SEL) [28]

process of TFT backplanes and OLEDs. In addition, the glass substrate plays the role of preventing size change during the TFT fabrication processes. On this substrate, TFT and OLED layers are fabricated. After encapsulating the active-matrix OLED devices, the glass substrate is finally de-bonded. Two common de-bonding methods are mechanical delamination [26] and laser delamination [27].

A bonding/de-bonding method has also been reported. In this method, a plastic film is bonded to a glass substrate as the first process. The later process is same as the coating/de-bonding method.

By contrast, the transfer method [28] has also been reported. The process flow is schematically illustrated in Fig. 10.12, in which the device structure is top emission OLED with a color filter. In this method, an inorganic separation layer, a passivation layer, a TFT layer, and an OLED layer are formed on a glass substrate. On a counter glass substrate, on the other hand, an inorganic separation layer, a passivation layer, and a color filter layer are formed. In the next step, the TFT/OLED substrate and the color filter substrate are bonded. After bonding, the glass substrates are separated from the passivation layers by physical force. Finally, flexible substrates are bonded to the AM-OLED device.

Some examples of flexible AM-OLED displays with a plastic film are summarized in Table 10.3.

Toshiba developed an 11.7″ flexible AM-OLED display with a polyimide film of 100 μm thickness, using the coat/de-bond method [29]. The display has qHD format (960 × 540 dots) with 94 ppi. They utilized a-IGZO TFTs with two transistors and one capacitor (2Tr + 1C), white emission with color filters (WOLED + CF), and bottom emitting device structure.

Sony developed a 9.9″ flexible OLED driven by IGZO TFTs [30]. They used the coat/de-bond method, where a flexible substrate with a high heat resistance of over 300 °C was directly formed on a glass substrate by coating. On the substrate, IGZO TFTs with two transistor circuits were fabricated. The field-effect mobility (μ) was 13.4 cm²/Vs. The device structure of the OLED is top emission combined with a color filter substrate having RGBW pixels. The flexible OLED display has qHD format (960 × 540 dots), a resolution of 111 ppi, a color gamut of 106%.

Table 10.3 Examples of flexible active-matrix OLED displays with a plastic film

Affiliation[*]	Size (diagonal)	Format	Resolution	TFT	RGB	Emitting direction	Substrate	Process	Comments	Ref.
Toshiba	11.7"	qHD	94 ppi	a-IGZO	White/CF	Bottom	PI	Coating/De-bonding		[29]
Sony	9.9"	qHD	111 ppi	IGZO	White/CF	Top	PI	Coating/De-bonding	106%NTSC	[30]
SEL	5.9	720×1280	249 ppi	CAAC-IGZO	White/CF	Top	PI	Transfer	Foldable	[28]
SEL	5.2	Quad VGA	302 ppi	CAAC-IGZO	White/CF (RGBW)	Top	PI	Transfer	Side-roll, top-roll	[28]
SEL	13.5"	8K4K	664 ppi	CAAC-IGZO	White/CF	Top		Transfer	Foldable	[31]
SEL	81	8K4K		CAAC-IGZO	White/CF	Top		Transfer	Kawara	[32]
Samsung	5.7	FHD	388 ppi						Commercialized	[33]
LG	5.97		245 ppi	ELA-TFT	RGB	Top	PI	Coating/De-bonding (by laser)	Commercialized	[27]
LG	18	WXGA		IGZO	RGB	Top	PI	Coating/De-bonding (by laser)		[34]
AUO	4.3′	qHD		LTPS		Top	PI	De-bonding (Mechanical)		[26]
Holst Centre	6 cm	QQVGA	85 ppi	IGZO solution	Monochrome	Top	PI	De-bonding		[35]
Holst Centre			200 ppi	IGZO		Top	PEN	Bonding/ lamination		[36]
BOE	9.55"	640×432		a-IGZO	FMM	Top	PI	De-bonding		[37]
NHK[†]	8	VGA	100 ppi	IGZO	RGB side-by-side	Bottom	PEN	Bond/De-bonding		[38]
Plastic Logic	3.86"	100 ppi		OTFT	Mono-color	Bottom	PET etc.			[39]

[*] Affiliation of the first author
[†] Japan Broadcasting Corporation (Japan)
FMM = fine metal mask technology

In 2013, both Samsung [33] and LG Display [27] separately commercialized flexible AM-OLED displays. Their displays have a curved shape.

The specifications of the Samsung's curved OLED display are 5.7″ diagonal, pixels 1920×1080, resolution 388 ppi, and curvature radius of 400 mm [33].

The specifications of the flexible OLED display commercialized by LG Display [27] are size of 5.98″ diagonal, pixel number 720×1280, resolution 245 ppi, ELA-TFT on a PI backplane, radius of curvature 700 mm.

Semiconductor Energy Laboratory (SEL) developed various types of flexible AM-OLED displays, using the transfer method [28, 31, 32]. Some of them have large size such as 13.5″, high resolution such as over 664 ppi, side roll, top roll, foldable function, and Kawara-type. Figure 10.13 shows an 81″ flexible AM-OLED display using Kawara-type multi-display technology [32].

Yoon et al. of LG Display developed the world's first large size 18″ flexible OLED display [34]. Pictures of the developed flexible 18″ flexible OLED display are shown in Fig. 10.14.

For applying various plastic films with low temperature stability, low temperature TFT technologies have also been developed. Fruehauf et al. of University of Stuttgart (Germany)

Figure 10.13 Picture of the 81″ flexible AM-OLED display using Kawara-type multi-display technology [32]. (provided by Semiconductor Energy laboratory)
Display size: 81″ (Kawara-type multi-display with 36 panels (6×6) 13.5″ panels
Number of pixels: 7680×4320 (8 K UHD)
Resolution: 108 ppi
Color: Full color
Device structure: White tandem OLED (top emission) + color filter
Driving: Active-matrix with CAAC-IGZO TFT

Figure 10.14 Fabrication process of 18″ flexible OLED display reported by Yoon et al. [34]

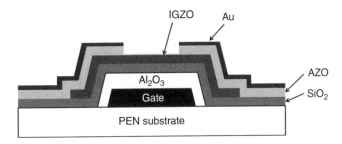

Figure 10.15 Typical cross-section of IGZO TFTs developed by Fruehauf et al. [40]. (Copyright 2015 The Japan Society of Applied Physics)

have reported that IGZO TFTs proves at the highest temperature of 160 °C [40]. Their device structure is shown in Fig. 10.15.

10.4 Flexible OLED Lighting

The addition of flexibility to OLED lighting can supply not only unique lighting products but also business competitiveness against competitors such as LEDs, providing several advantages to consumers. Indeed, flexibility can realize several excellent features such as small thickness, light weight, and design flexibility, which are not easily achieved by previous lighting technologies including LED. In addition, flexibility can produce a production revolution by roll-to-roll (R2R) processes, which can induce a drastic reduction in production cost.

Three types of flexible substrates (ultra-thin glass, stainless steel foil, and plastic film), which are described in Section 10.2, can be applied to flexible OLED lighting.

10.4.1   Flexible OLED Lighting on Ultra-Thin Glass

In 2013, LG Chem (South Korea) commercialized the world's first flexible OLED lighting panels fabricated on panels of ultra-thin glass [41, 42]. The specifications are 210×50 mm, 55 lm/W power efficiency, and 4000 K in color temperature. The panels are 0.2 mm thick and weigh just 0.6 g. The OLED devices have a hybrid structure (a combination of phosphorescent plus fluorescent emitters).

In addition, further developments of flexible OLEDs fabricated using ultra-thin glass are continuing.

NEC Lighting (Japan) developed a 9.2×9.2 cm OLED panel using ultra-thin glass, being encapsulated by stainless steel foil [12]. The utilized ultra-thin glass was developed by Nippon Electric Glass [6]. By using encapsulation with stainless steel foil, mechanical reliability increases. The device structure and the picture of emission are shown in Fig. 10.16.

Furukawa et al. reported a flexible OLED device with an ultra-thin glass developed in a collaboration between Yamagata University and the seven companies in Yamagata University

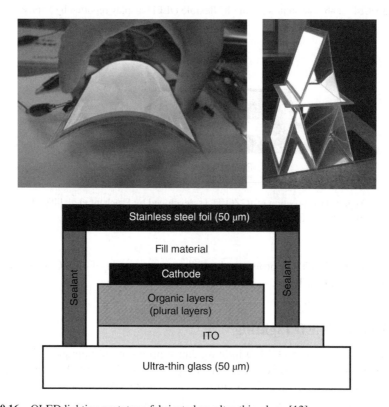

Figure 10.16   OLED lighting prototype fabricated on ultra-thin glass. [12]
Substrate size: 50×50 mm
Emission area: 32×32 mm
Panel fabrication: NEC Lighting Ltd
Ultra-thin glass: developed by Nippon Electric Glass [6]
Stainless steel foil: developed by Nippon Steel and Sumitomo Metal Corporation Group [11]

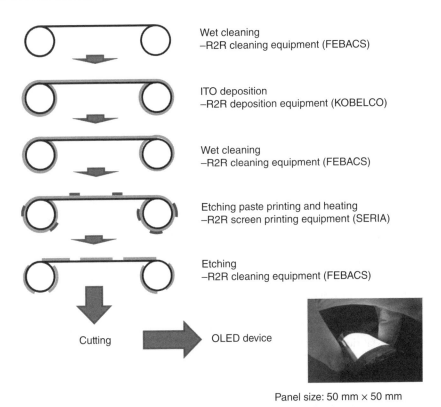

Wet cleaning
–R2R cleaning equipment (FEBACS)

ITO deposition
–R2R deposition equipment (KOBELCO)

Wet cleaning
–R2R cleaning equipment (FEBACS)

Etching paste printing and heating
–R2R screen printing equipment (SERIA)

Etching
–R2R cleaning equipment (FEBACS)

Cutting → OLED device

Panel size: 50 mm × 50 mm

Emission area: 32 mm × 32 mm

Figure 10.17   The process flow for fabricating ITO patterns on ultra-thin glass and a flexible OLED device using the fabricated ultra-thin glass with patterned ITO [43]

Organic Thin Film Device Consortium [43]. They used a roll glass with the thickness of 50 μm and width of 300 mm. The process flow for fabricating ITO patterns is shown in Fig. 10.17. After the roll of ultra-thin glass is washed by the roll-to-roll wet cleaning equipment developed by Febacs (Japan), an ITO layer is deposited by using the roll-to-roll deposition equipment developed by Kobe Steel (Japan). The sputtering ITO target contains 10 wt% of SnO_2. It was reported that the ITO film deposited at −20 °C showed very low membrane stress, small of crystal size and flat surface, compared with the ITO film deposited at 250 °C. The resistance and the transmittance of the ITO film deposited at −20 °C were 18 Ω/sq and 85%, respectively, after annealing at 250 °C. As a patterning technique of the ITO film, they used an etching paste instead of normal photolithography technique. In their process, after the wet cleaning of the ultra-thin glass with the deposited ITO film by the roll-to-roll wet cleaning equipment, an etching paste was printed by the roll-to-roll screen printing equipment developed by Tokai Shoji (Japan) [44]. The etching paste dissolved the ITO film by heat treatment. After the etching of the ITO, the ultra-thin glass was washed again by the roll-to-roll wet cleaning equipment. Using this ultra-thin glass, they fabricated an OLED device with the emission area 32 mm square as shown in Fig. 10.17.

10.4.2   Flexible OLED Lighting on Stainless Steel Foil

Since stainless steel foil is not transparent, OLED fabricated on stainless steel foil has to have a top emission device structure.

Ma et al. of Universal Display Corporation (USA) developed flexible OLED lighting panels with a size of $15 \, cm \times 15 \, cm$, using a stainless steel foil with the thickness of $30 \, \mu m$ and phosphorescent OLED [45]. For planarization of the surface of stainless steel foil, they coated a thermally cured polyimide by the spin-coating method. For reducing the resistivity of the anode, Al bus lines were used.

Yamada et al. reported flexible OLED lighting with stainless steel foil developed by the collaboration of Yamagata University with Nippon Steel and Sumitomo Metal Corporation Group [12, 46]. The device structure, the V–I characteristics, and a picture of the emission are shown in Fig. 10.18. The stainless steel foil has an inorganic–organic hybrid insulating layer of $3 \, \mu m$ [11]. As is obvious in Fig. 10.18, the V–I characteristics show low leakage current before the turn-on voltage. This means that the insulating layer effectively eliminates the cause of electrical shorts or leakage between the anode and the cathode. They successfully fabricated OLED lighting panels with an emission area of $32 \times 32 \, mm$ by using stainless steel foils with the inorganic–organic hybrid insulating layer.

They also successfully fabricated OLED devices with an emission area of $75 \times 75 \, mm$, using the coated stainless steel foil. NEC lighting (Japan) fabricated a top emitting OLED device structure fabricated on this stainless steel foil. The device structure and an emission picture are shown in Fig. 10.19 [12].

10.4.3   Flexible OLED Lighting on Plastic Films

Ma et al. of Universal Display Corporation (USA) developed flexible OLED lighting panels with a size of $15 \times 15 \, cm$, using a planarized PEN substrate, phosphorescent OLED, and an out-coupling film [45]. The specifications are luminance $3000 \, cd/m^2$, efficiency $43 \, lm/W$, CRI 84, 1931 CIE of (0.435, 0.426), and CCT $3200 \, K$.

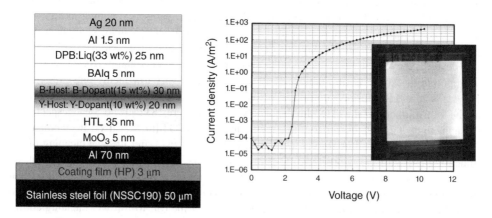

Figure 10.18   Device structure, the V-I characteristics and a picture of emission of the OLED device fabricated on a stainless steel foil [12, 46]

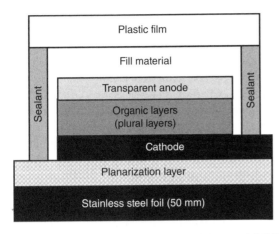

Figure 10.19   An OLED device on a stainless steel foil [12]
Panel fabrication: NEC Lighting
Stainless steel foil: Nippon Steel and Sumitomo Metal Corporation Group
Panel size: 92×92 mm
Emission area: 75×75 mm

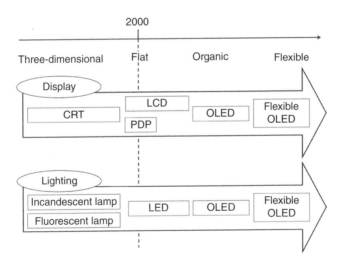

Figure 10.20   The generation change of displays and lighting

Ohsawa et al. of Advanced Film Device Inc. (Japan) and Semiconductor Energy Laboratory (Japan) developed a flexible OLED lighting prototype with 360×360 mm, using the transfer method [47]. The device was encapsulated by a stainless steel foil. The specifications are luminance 1000 cd/m², power efficiency 111 lm/W, CIE (0.49, 0.50), and CCT 2860 K.

Tsujimura et al. of Konica Minolta (Japan) announced that flexible OLED panels are being manufactured with the world's first roll-to-roll equipment for OLEDs using plastic barrier film [16]. These panels were commercialized in 2014. They utilized an excellent gas barrier film with WVTR $<5.9 \times 10^{-5}$ g/m²/day and reported that no dark spot appeared after 300 hours at 85 °C/85%RH.

10.5   Toward the Flexible

Figure 10.20 shows the generation changes of displays and lightings. The classic displays and lighting devices have had three-dimensional shapes such as CRT (cathode ray tube), incandescent lamps, and fluorescent lamps. Since the end of the 20th century, they have been changing because of flat shaped devices such as LCD (liquid crystal display), PDP (plasma display panel), EL (electroluminescent inorganic device), and flat lighting with LEDs. In particular, in display applications, the change to flat had a tremendous impact on society and life because flat panel displays made possible notebook PCs, mobile phones, smart phones, digital cameras with a display, video cameras with a display, and large size TVs. If there were no flat panel displays, we could not contact with our friends by email on a mobile phone and we could not access the internet, on-line games, or various networked communications by using smart phones. It might even be said that "no FPD (flat panel display), no life".

At present, OLEDs are aiming to replace the current flat shaped devices, although the business barrier is not low. On the other hand, we may believe that the next generation is going to be "flexible", due to the attractive features of flexible OLEDs and the current technological progress induced by the efforts of scientists and engineers. The realization of flexible OLEDs can induce the change from the current flat devices to flexible OLEDs. We can say that the challenge to flexible OLEDs will change our life and produce big business, as LCDs have done over the past 30 years.

References

[1] G. Gustafsson, Y. Cao, G. M. Treacy, F. Klavetter, N. Coleneri and A. J. Heeger, *Nature*, **357**, 477–479 (1992).

[2] G. Gu, P. E. Burrows, S. Venkatesh, S. R. Forrest, M. E. Thompson, *Opt. Lett.* **22**, 172 (1997).

[3] C. C. Wu, S. D. Theiss, G. Gu, M. H. Lu, J. C. Sturm, S. Wagner, S. R. Forrest, *SID 97 Digest*, 7.2 (p. 67) (1997).

[4] M. Nogi, S. Iwamoto, A. N. Nakagaito, H. Yano, *Adv. Mater.*, **21**, 1595–1598 (2009).

[5] G. Banzashi, H. Fushimi, S. Iwai, M. Tsunoda, E. Mikami, *Proc. IDW'14*, FLX5–3 (p. 1452) (2014).

[6] K. Fujiwara, *New Glass*, **24**, 90 (2009).

[7] Y. Matsuyama, K. Ebata, D. Uchida, T. Higuchi and S. Kondo, *Proc. IDW'13*, FLX2–1 (p. 1518) (2013).

[8] N. Inayama and T. Fujii, *Proc. IDW'13*, FLX4-4 (2013).

[9] Y. Ikari and H. Tamagaki, *Proc. IDW'12*, FLX5/FMC5-1 (p. 1493) (p. 1552) (2012).

[10] Z. Xie, L. S. Hung, F. Zhu, *Chem. Phys. Lett.*, **381**, 691–696 (2003).

[11] N. Yamada, T. Ogura, S. Ito and K. Nose, *Proc. IDW'10*, FLXp-5 (p. 2217) (2010); N. Yamada, T. Ogura, S. Ito, and K. Nose, *Proc. IDW'11*, FLX6-2 (p. 2013) (2011).

[12] M. Koden, H. Kobayashi, T. Moriya, N. Kawamura, T. Furukawa and H. Nakada, *Proc. IDW'14*, FLX6/FMC-1 (p. 1454) (2014); M. Koden, *Proc. The Twenty-second International Workshop on Active-matrix Flatpanel Displays and Devices (AM-FPD 15)*, 2–1 (p. 13) (2015).

[13] P. E. Burrows, G. L. Graff, M. E. Gross, P. M. Martin, M. K. Shi, M. Hall, E. Mast, C. Bonham, W. Bennett, M. B. Sullivan, *Displays*, **22**, 65–69 (2001).

[14] M. S. Weaver, L. A. Michalski, K. Rajan, M. A. Rothman, J. A. Silvernail J. J. Brown, P. E. Burrows, G. L. Graff, M. E. Gross, P. M. Martin, M. Hall, E. Mast, C. Bonham, W. Bennett, M. Zumhoff, *Appl. Phys. Lett.*, **81(16)**, 2929–2931 (2002).

[15] Y. Suzuki, K. Nishijima, S. Naganawa, K. Nagamoto, T. Kondo, *SID 2014 Digest*, 6.4 (p. 56) (2014).

[16] T. Tsujimura, J. Fukawa, K. Endoh, Y. Suzuki, K. Hirabayashi, T. Mori, *SID 2014 Digest*, 10.1 (p. 104) (2014).

[17] M. D. Groner, S. M. George, R. S. McLean, P. F. Carcia, *Appl. Phys. Lett.*, **88**, 051907 (2006).

[18] T. Miyake, A. Yoshida, T. Yoshizawa, A. Sugimoto, H. Kubota, T. Miyadera, M. Tsuchida, H. Nakada, *Proc. of IDW'03*, OEL2–2 (p. 1289) (2003).

[19] A. Sugimoto, A. Yoshida, T. Miyadera, *Technical Report of Pioneer R&D*, **11(3)**, 48–56; T. Nagashima, H. Yamada, M. Hanaoka, T. Ichikawa, T. Ishida, K. Oda, *Pioneer R&D*, **13(3)**, 65–73.

[20] C. C. Kuo, J.-Y. Chiou, S.-F. Liu, C.-H. Chiu, C.-H. Lin, Y.-C. Sun, M.-C.-Chen, Y.-W Chiu, *Proc. IDW'13*, OLED3–3 (p. 886) (2013).

[21] S. D. Theiss, S. Wagner, *IEEE Electron Dev. Lett.*, **17(12)**, 578–580 (1996).

[22] M. N. Troccoli, A. J. Roudbari, T.-K. Chuang, M. K. Hatalis, *Solid-State Electronics*, **50**, 1080–1087 (2006).

[23] D. U. Jin, J. K. Jeong, H. S. Shin, M. K. Kim, T. K. Ahn, S. Y. Kwon, J. H. Kwack, T. W. Kim, Y. G. Mo, H. K. Chung, *SID 06 Digest*, 64.1 (p. 1855) (2006).

[24] F. Templier, B. Aventurier, P. Demars, J.-L. Botrel, P. Martin, *Thin Solid Films*, **515**, 7428–7432 (2007).

[25] R.-Q. Ma, K. Rajan, M. Hack, J. J. Brown, J. H. Cheon, S. H. Kim, M. H. Kang, W. G. Lee, J. Jang, *SID 08 Digest*, 30.3 (p. 425) (2008).

[26] Y.-L. Lin, T.-Y. Ke, C.-J. Liu, C.-S. Huang, P.-Y. Lin, C.-H. Tsai, C.-H. Tu, P.-F. Wang, H.-H. Lu, M.-T. Lee, K.-L. Hwu, C.-S. Chuang, Y.-H. Lin, *SID 2014 Digest*, 10.4 (p. 114) (2014).

[27] S. Hong, C. Jeon, S. Song, J. Kim, J. Lee, D. Kim, S. Jeong, H. Nam, J. Lee, W. Yang, S. Park, Y. Tak, J. Ryu, C. Kim, B. Ahn, S. Yeo, *SID 2014 Digest*, 25.4 (p. 334) (2014).

[28] R. Kataish, T. Sasaki, K. Toyotaka, H. Miyake, Y. Yanagisawa, H. Ikeda, H. Nakashima, N. Ohsawa, S. Eguchi, S. Seo, Y. Hirakata, S. Yamazaki, C. Bower, D. Cotton, A. Matthews, P. Andrew, C. Gheorghiu, J. Bergquist, *SID 2014 Digest*, 15.3 (p. 187) (2014); Y. Jimbo, T. Aoyama, N. Ohno, S. Eguchi, S. Kawashima, H. Ikeda, Y. Hirakata, S. Yamazaki, M. Nakada, M. Sato, S. Yasumoto, C. Bower, D. Cotton, A. Matthews, P. Andrew, C. Gheorghiu, J. Bergquist, *SID 2014 Digest*, 25.1 (p. 322) (2014); R. Komatsu, R. Nakazato, T. Sasaki, A. Suzuki, N. Senda, T. Kawata, H. Ikeda, S. Eguchi, Y. Hirakata, S. Yamazaki, T. Shiraishi, S. Yasumoto, C. Bower, D. Cotton, A. Matthews, P. Andrew, C. Gheorghiu, J. Bergquist, *SID 2014 Digest*, 25.2 (p. 326) (2014); J. Koezuka, K. Okazaki, S. Idojiri, Y. Shima, K. Takahashi, D. Nakamura, S. Yamazaki, *Proc. AM-FPD'15*, 4–1 (p. 205) (2015).

[29] H. Yamaguchi, T. Ueda, K. Miura, N. Saito, S. Nakano, T. Sakano, K. Sugi, I. Amemiya, M. Hiramatsu, A. Ishida, *SID 2012 Digest*, 74.2 L (p. 1002) (2012); H. Yamaguchi, T. Ueda, K. Miura, N. Saito, S. Nakano, T. Sakano, K. Sugi, I. Amemiya, *Proc. IDW/AD'12*, AMD8/FLX7–1 (p. 851) (2012).

[30] M. Noda, K. Teramoto, E. Fukumoto, T. Fukuda, K. Shimokawa, T. Saito, T. Tanikawa, M. Suzuki, G. Izumi, S. Kumon, T. Arai, T. Kamei, M. Kodate, S. No, T. Sasaoka, K. Nomoto, *SID 2012 Digest*, 74.1 L (p. 998) (2012); K. Teramoto, E. Fukumoto, T. Fukuda, K. Shimokawa, T. Saito, T. Tanikawa, M. Suzuki, G. Izumi, M. Noda, S. Kumon, T. Arai, T. Kamei, M. Kodate, S. No, T. Sasaoka, K. Nomoto, *Proc. IDW/AD'12*, AMD8/FLX7–2 (p. 855) (2012).

[31] K. Takahashi, T. Sato, R. Yamamoto, H. Shishido, T. Isa, S. Eguchi, H. Miyake, Y. Hirakata, S. Yamazaki, R. Sato, H. Matsumoto, N. Yazaki, *SID 2015 Digest*, 18.4 (p. 250) (2015).

[32] D. Nakamura, H. Ikeda, N. Sugisawa, Y. Yanagisawa, S. Eguchi, S. Kawashima, M. Shiokawa, H. Miyake, Y. Hirakata, S. Yamazaki, S. Idojiri, A. Ishii, M. Yokoyama, *SID 2015 Digest*, 70.2 (p. 1031) (2015).

[33] News release of Samsung Electronics, 9 October 2013: http://global.samsungtomorrow.com/?p=28863.

[34] J. Yoon, H. Kwon, M. Lee, Y.-Y. Yu, N. Cheong, S. Min, J. Choi, H. Im, K. Lee, J. Jo, H. Kim, H. Choi, Y. Lee, C. Yoo, S. Kuk, M. Cho, S. Kwon, W. Park, S. Yoon, I. Kang, S. Yeo, *SID 2015 Digest*, 65.1 (p. 962) (2015).

[35] B. Cobb, F. G. Rodriguez, J. Maas, T. Ellis, J.-L. van der Steen, K. Myny, S. Smout, P. Vicca, A. Bhoolokam, M. Rockelé, S. Steudel, P. Heremans, M. Marinkovic, D.-V. Pham, A. Hoppe, J. Steiger, R. Anselmann, G. Gelinck, *SID 2014 Digest*, 13.4 (p. 161) (2014).

[36] F. Li, E. Smits, L. van Leuken, G. de Haas, T. Ellis, J.-L. van der Steen, A. Tripathi, K. Myny, M. Ameys, S. Schols, P. Heremans, G. Gelinck, *SID 2014 Digest*, 32.2 (p. 431) (2014).

[37] S. Shi, D. Wang, J. Yang, W. Zhou, Y. Li, T. Sun, K. Nagayama, *SID 2014 Digest*, 25.3 (p. 330) (2014).

[38] H. Fukagawa, K. Morii, M. Hasegawa, Y. Nakajima, T. Takei, G. Motomura, H. Tsuji, M. Nakata, Y. Fujisaki, T. Shimizu, T. Yamamoto, *SID 2014 Digest*, P-154 (p. 1561) (2014).

[39] C. W. M. Harrison, D. K. Garden, I. P. Horne, *SID 2014 Digest*, 20.3L (p. 256) (2014).

[40] N. Fruehauf, M. Herrmann, H. Baur, M. Aman, *Proc. AM-FPD'15*, S1–2 (p. 39) (2015).

[41] LG Chem, Press release, 3 April 2013: www.lgchem.com/global/lg-chem-company/information-center/press-release/news-detail-527

[42] LG Chem, Press release, 30 September 2013: www.lgchem.com/global/lg-chem-company/information-center/press-release/news-detail-567

[43] T. Furukawa, K. Mitsugi, H. Itoh, D. Kobayashi, T. Suzuki, H. Kuroiwa, M. Sakakibara, K. Tanaka, N. Kawamura, M. Koden, *Proc. IDW'14*, FLX4-3 L (p. 1428) (2014).

[44] D. Kobayashi, N. Naoi, T. Suzuki, T. Sasaki, T. Furukawa, *Proc. IDW'14*, FLX3-1 (p. 1417) (2014).

[45] R. Ma, H. Pang, P. Mandlik, P. A. Levermore, K. Rajan, J. Silvernail, E. Krall, J. Paynter, M. Hack, J. J. Brown, *SID 2012 Digest*, 57.1 (p. 772) (2012).

[46] N. Yamada, H. Kobayashi, S. Yamaguchi, J. Nakatsuka, K. Nose, K. Uemura, M. Koden, H. Nakada, *Proc. IDW'14*, FLX6/FMC6-4L (p. 1465) (2014).

[47] N. Ohsawa, S. Idojiri, K. Kumakura, S. Obana, Y. Kobayashi, M. Kataniwa, T. Ohide, M. Ohno, H. Adachi, N. Sakamoto, S. Yatsuzuka, T. Aoyama, S. Yamazaki, *SID 2013 Digest*, 66.4 (p. 923) (2013).

[48] K Furukawa, K. Kato, T. Iwasaki, *Proc. IDW'13*, OLED5-1 (p. 902) (2013); K. Hirabayashi, H. Ito, T. Mori, *Proc. IDW'13*, FLX4-2 (p. 1546) (2013).

11

New Technologies

Summary

New technologies generate industrial innovations. This chapter describes several new technologies: non-ITO transparent electrode, organic TFT, wet-processed TFT, wet-processed OLED, roll-to-roll equipment, and quantum dot. Most of these are compatible with wet processes and are flexible. This means that wet processes and flexible are strongly required and related to the generation change of organic electronics technologies because wet processes and flexible give new value to products and low cost.

Key words

non-ITO transparent electrode, conducting polymer, silver nanowire, CNT, printing, roll-to-roll, organic TFT, wet process, quantum dot

11.1 Non-ITO Transparent Electrodes

ITO (indium-tin-oxide) is the most frequently used transparent electrode not only in OLEDs but also in LCDs. However, ITO has several issues, especially in flexible devices. First, mechanical flexibility of ITO is limited; it is brittle in bending, generating cracks and defects. Second, fabrication of ITO film usually requires long process steps including vacuum deposition process and photolithography, which represent a significant investment. Third, indium is a rare metal, inducing the issue of stable supply due to the depletion of resources. For the latter two reasons ITO is expensive.

Therefore, alternative electrodes to ITO are strongly demanded and this has become one of the hottest recent trends. Alternative candidates are conducting polymer [1–10], stacked layer using Ag [11–16], silver nanowire (AgNW) [4–6, 17–19], carbon nanotube (CNT) [20–29],

OLED Displays and Lighting, First Edition. Mitsuhiro Koden.
© 2017 John Wiley & Sons, Ltd. Published 2017 by John Wiley & Sons, Ltd.

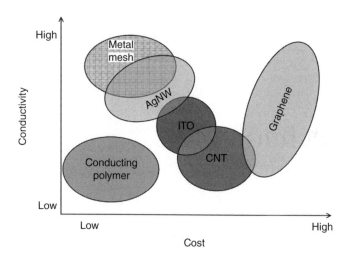

Figure 11.1   An example illustrating the relationship between cost and conductivity of various transparent electrode materials. (Source: ref. [30])

metal mesh, and graphene. Figure 11.1 shows an example illustrating the relationship between costs and conductivities of various transparent electrode materials [30].

While conducting polymer has much lower conductivity than ITO, the attractive point is its low cost. Non-ITO transparent materials using Ag are estimated to be able to achieve lower cost and higher conductivity than ITO. On the other hand, carbon materials such as CNT and graphene still have issues of cost and conductivity.

11.1.1   Conducting Polymer

One of the most promising conducting polymers seems to be poly(3,4-ethylenedioxythiophene):poly(styrene-sulfonate) PEDOT:PSS and their analogs and derivatives. The molecular structure of PEDOT:PSS is shown in Fig. 4.36. While PEDOT:PSS can be used as a hole injection and/or hole transport material in OLEDs, it is also attractive as an electrode material.

The conductivity of PEDOT:PSS itself is not very high. For example, the conductivity of Baytron P, which is a frequently used PEDOT:PSS material supplied from Bayer Corporation, is less than 10 S/cm. However, the conductivity of PEDOT:PSS can be drastically enhanced by adding certain compounds such as poly-alcohols (alcohols with more than two OH group on each molecule) [1], high-dielectric solvents such as dimethyl sulfoxide (DMSO) [2], and N,N-dimethylformamide (DMAc) [3].

Kim et al. of the US Naval Research Laboratory reported OLED devices with PEDOT:PSS containing a small amount of glycerol, obtaining comparable performance with OLED devices with ITO [1].

Ouyang et al. of the University of California, Los Angeles (USA) reported that the conductivity of PEDOT:PSS was improved to 160 S/cm, by adding ethylene glycol or *meso*-erythritol [2]. They fabricated polymer OLED devices by using the highly conductive PEDOT:PSS, obtaining a performance close to that of an OLED device with ITO instead of PEDOT:PSS.

Table 11.1 Phosphorescent OLED devices with PEDOT:PSS electrodes [3]

Color	Electrode	Voltage (V)	Current density (mA/cm^2)	Current efficiency (cd/A)	Power efficiency (lm/A)	CIE coordinate
Green	PEDOT:PSS	3.07±0.06	0.16±0.02	62.0±2.3	63.5±3.3	(0.30, 0.64)
	ITO	3.16±0.05	0.19±0.03	54.1±1.1	53.8±1.9	(0.32, 0.62)
Blue	PEDOT:PSS	3.25±0.06	2.35±0.05	4.2±0.1	4.0±0.1	(0.16, 0.24)
	ITO	3.51±0.04	2.7±0.05	3.7±0.1	3.3±0.1	(0.16, 0.22)
Red	PEDOT:PSS	2.61±0.02	0.75±0.1	13.2±3.0	15.9±3.1	(0.65, 0.36)
	ITO	2.64±0.03	0.7±0.1	14.6±3.3	17.3±4.1	(0.64, 0.35)

Luminance of OLED devices: 100 cd/m^2.

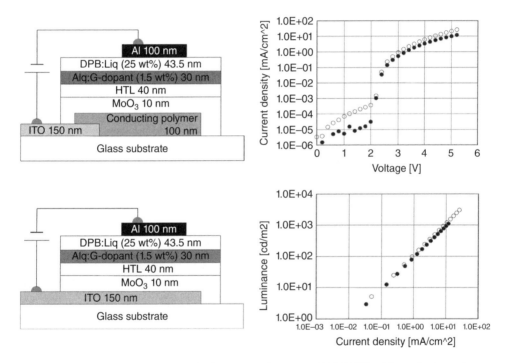

Figure 11.2 Device structure and typical OLED performance of (a) OLED with conducting polymer (closed circle) and (b) OLED with ITO (open circle) [4–6]

Fehse et al. of Technische Universität Dresden and H. C. Starck GmbH (Germany) reported phosphorescent OLED devices with PEDOT:PSS electrodes instead of ITO [3]. The PEDOT:PSS is a Baytron PH 500 (H. C. Starck) containing 5% DMSO. The results are summarized in Table 11.1. The OLED devices with PEDOT:PSS shows better or comparable performance than those of corresponding OLED devices with ITO.

My colleagues and I at Yamagata University (Japan) reported on the application of a conducting polymer to OLED devices [4–6]. The device structure and typical performance are shown in Fig. 11.2, accompanying with those of the reference device with ITO. The conductivity of the developed conducting polymer is around 100 S/cm. As shown in Fig. 11.2, the OLED with the conducting polymer shows comparable I–V–L characteristics to the reference OLED with ITO. The driving voltage of the OLED with the conducting polymer is slightly higher than that of the OLED with ITO. This is caused by the difference in the conductivity between the conducting polymer and ITO. These results indicate that the conducting polymer can be used in OLED, giving no serious damage to OLED properties.

We also reported a 5×5 cm OLED panel using the conducting polymer, as shown in Fig. 11.3 [4, 5]. The OLED panel has metal bus lines with 1.5 mm pitch in order to reduce electrical resistance. The pitch was decided by the OLED luminance uniformity simulation based on the conductivity of the conducting polymer. The picture indicates that an uniform emission was obtained in spite of the absence of ITO.

Patterning technologies of conducting polymers have also been studied and reported. Piliego et al. of National Nanotechnology Laboratory of INFM-CNR, and Istituto Superiore Universitario di Formazione Interdisciplinare-sezione Nanoscienze (Italy) reported a lift-up soft lithography technique to pattern a highly conductive PEDOT-PSS [7]. They spin-coated an aqueous polymer dispersion of Baytron PH500 containing 5% of DMSO onto a pre-cleaned glass substrate. The thickness of the PEDOT:PSS film is almost 100 nm and conductivity is about 360 S/cm. They also prepared a mold by casting polydimethylsiloxane (PDMS) (Sylgard 184) with a curing agent in the ratio of 10:1 against a silicon template realized by photolithography. The mold is contacted with the PEDOT:PSS film. When the mold is removed, the PEDOT:PSS layer contacting the mold is eliminated. They fabricated 55 μm lines of PEDOT-PH500 separated by 70 μm, 3 μm lines separated by 15 μm, 15 μm lines separated by 15 μm, square holes of size 60×60 μm and spacing 15 μm, etc. By using such PEDOT:PSS patterns, ITO-free OLED devices were reported to be successfully fabricated.

(a)   (b)

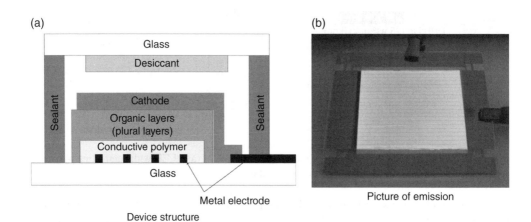

Device structure   Picture of emission

Figure 11.3 OLED panel using the conducting polymer. [4, 5] (a) Device structure. (b) A picture of emission
*Substrate size: 50×50 mm
*Emission aria: 32×32 mm
*Conductive polymer: spin-coated
*Metal electrode: Mo/Al/Mo (pitch 1.5 mm, width 30 μm, thickness 400 nm)

Ha et al. of Seoul National University (Korea) reported PEDOT-PSS anode deposited by ink-jet printing [8]. As the anode material, they used PEDOT-PSS (E-157) supplied from Con-Tech Co., Ltd (Korea). They achieved 125 Ω/sq with a transmittance of 85.8% at 500 nm, by four times ink-jet printing of the PEDOT-PSS (E-157). The work function of the PEDOT-PSS film is 5.34 eV. They applied the PEDOT-PSS (E-157) film to an OLED with the structure of glass, PEDOT:PSS (E-157) as anode, PEDOT:PSS (AI 4083) as hole injection layer (HIL), emission layer (EML, SPG-01 T from Merck, Germany), LiF as electron injection layer (EIL), and Al. The device showed the same turn-on voltage and a slightly lower current efficiency, comparing with the reference device with ITO instead of the PEDOT-PSS (E-157) film.

Ouyang et al. of Shanghai Jiao Tong University (China) reported on the photolithographic patterning of PEDOT:PSS, using a silver interlayer on the PEDOT:PSS [9, 10]. The silver is patterned by wet etching, and the PEDOT:PSS is patterned by dry etching with oxygen plasma. They reported that the conductivity of the PEDOT:PSS was nearly unchanged in the patterning process. They fabricated a PEDOT:PSS pattern, in which the line width is 20 μm and the gap between lines is also 20 μm. They also fabricated OLED devices with the developed PEDOT:PSS patterns and found that the efficiencies of the OLED devices are comparable with an OLED device with ITO instead of PEDOT:PSS.

Furukawa et al. of Yamagata University (Japan) developed a flexography printing of conducting polymers [5]. By optimizing printing ink and trench shape of printing plate, they successfully printed a conducting polymer with the desired pattern. Using the developed printing technology, they fabricated an OLED device as shown in Fig. 11.4. In the OLED device, the bottom electrode (anode) consisted of conducting polymer printed by flexography

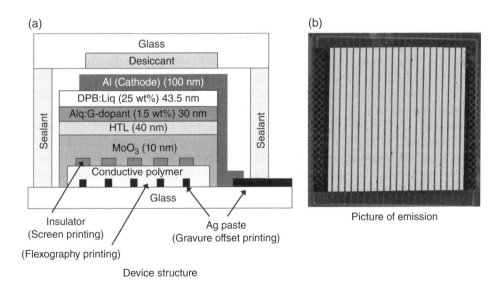

Figure 11.4 OLED panel with conducting polymer printed by flexography printing, stripe assisting Ag electrode printed by gravure offset printing, and insulating pattern printed by screen printing. (a) Device structure. (b) A picture of emission
*Substrate size: 50×50 mm
*Emission aria: 32×32 mm

printing, stripe assisting Ag electrode printed by gravure offset printing and insulating pattern printed by screen printing.

11.1.2 Stacked Layer Using Ag

Silver is transparent when it is thinner than about 30 nm. However, Ag itself has several issues such as oxidization, migration, and low work function (low hole injection ability). As the counter technology, stacking layers using Ag have been investigated. Some examples are summarized in Table 11.2.

Choi et al. at Seoul National University and Samsung Display Devices (Korea) reported on an ITO/Ag/ITO electrode, while they were aiming at the application to STN (super twisted nematic)-LCDs [11]. According to their report, an ITO/Ag/ITO multi-layer structure with Ag thickness of 14 nm and the ITO thickness of 55–60 nm showed a low sheet resistance of 4 Ω/sq and high optical transmittance of 90% at 550 nm.

Fahland et al. [12] of Fraunhofer Institute Electron Beam and Plasma Technology (FEP) (Germany) deposited ITO/Ag/ITO electrodes on a polyethylentherephtalate (PET) film with a thickness of 75 μm by a roll-to-roll sputtering equipment, achieving sheet resistance of lower than 16 Ω/sq and transmittance of 80% at 550 nm. Typical thicknesses were 50 nm for ITO and 8 nm for Ag.

Li et al. in Fuzhou University (People's Republic of China) reported on the application of Al-doped zinc oxide (AZO)/Ag/Al-doped zinc oxide (AZO) anode to OLED devices [15]. AZO is one of candidates for non-ITO materials due to the several advantages such as high transmittance, abundant resource in nature, and non-toxicity. However, AZO thin films have a higher resistivity compared with ITO thin films. In order to solve this problem, they investigated a stacked AZO(40 nm)/Ag(12 nm)/AZO(40 nm) structure. The electrode showed an excellent low resistance of 8.3 Ω/sq and a transmittance of over 80% in wavelength range 400–800 nm. Using this electrode, they fabricated an OLED device with the structure of glass/AZO(40 nm)/Ag(12 nm)/AZO(40 nm)/TPD(100 nm)/Alq$_3$(60 nm)/LiF(0.5 nm)/Al(200 nm), obtaining a better efficiency (4.97 cd/A) than that of a reference OLED device with an ITO electrode (2.98 cd/A).

Han et al. in KAIST (Republic of Korea) reported on the application of ZnS/Ag/MoO$_3$ multi-layer anode to OLEDs [16]. They fabricated a ZnS(25 nm)/Ag(7 nm)/MoO$_3$(5 nm) multi-layer anode, obtaining a transmittance of 83% at λ=550 nm and a sheet resistance of 9.6 Ω/sq. They reported that the sheet resistance of ZnS/Ag/MoO$_3$ multi-layer anode does not change by bending of flexible substrates, while the sheet resistance of ITO changes

Table 11.2 Examples of stacking layers using Ag

Structure	Sheet resistance	Transmittance	Ref.
ITO(55–60 nm)/Ag(14 nm)/ITO(55–60 nm)	4 Ω/sq	90% (550 nm)	[11]
ITO(50 nm)/Ag(8 nm)/ITO(50 nm)	16 Ω/sq	80% (550 nm)	[12]
ITO(50 nm)/Ag(8 nm)/ITO(50 nm)	23 Ω/sq	80.3% (max in 400–600 nm)	[13]
ITO(60 nm)/Ag(10 nm)/ITO(60 nm)	6 Ω/sq	80% (550 nm)	[14]
AZO(40 nm)/Ag(12 nm)/AZO(40 nm)	8.3 Ω/sq	Over 80% (400 nm–800 nm)	[15]
ZnS(25 nm)/Ag(7 nm)/MoO$_3$ (5 nm)	9.6 Ω/sq	83% (550 nm)	[16]

considerably by the bending of flexible substrates. They also reported that a flexible OLED device with the multi-layer anode on a PEN substrate showed feasible I–L–V characteristics compared with those of ITO-based one.

11.1.3   Silver Nanowire (AgNW)

Silver nanowire (AgNW) is a promising candidate for non-ITO transparent electrode materials.

Yu et al. of University of California (USA) and China University of Mining and Technology (P. R. China) reported on polymer OLED devices with silver nanowire electrode [17]. Their device structure was polyacrylate/AgNW/PEDOT:PSS/SY-PPV/CeF/Al.

Jiu et al. of Osaka University and Showa Denko (Japan) developed silver nanowire conductive film fabricated by a high-intensity pulsed light technique [18]. They reported to achieve sheet resistance of $19\,\Omega/\text{sq}$ and transmittance of 83% at 550 nm in their flexible AgNW film by only one step on a polyethylene terephthalate substrate with a light exposure with the intensity of $1.14\,\text{J/cm}^2$.

Pschenitzka of Cambrios Technologies Corporation (USA) reported the application of AgNW to OLED [19]. A nanowire suspension (Cambrios' ClearOhm™ product) was spin-coated on a glass substrate. The film was dried for 90 seconds at $50\,^\circ\text{C}$ and followed by a soft bake for 90 seconds at $140\,^\circ\text{C}$ on a hotplate. The sheet resistance of the nanowire film is reported to be $10\,\Omega/\text{sq}$ with a total transmission (including the glass substrate) of 86%. The standard photolithography and an acid-base etching were used for patterning the nanowire film. The patterned nanowire film was then treated with a vacuum Ar plasma process to remove organic material to expose the silver nanowire. On the nanowire film, OLED layers and a cathode were evaporated. He also investigated the influence of the thickness of hole-injection layer (HIL) which is directly deposited on the silver nanowire. When the thickness of HIL is 175 nm, sub-threshold leakage of the OLED devices is observed. On the other hand, when the thicknesses of HIL are 330 nm, 530 nm, and 800 nm, sub-threshold leakage is not observed. This result indicates that the surface roughness of the silver nanowire should be taken care of, and some covering and/or planarization technologies are required. He reported that $10 \times 10\,\text{cm}$ prototype white OLED was successfully fabricated by using the silver nanowire film. He also reported the better lifetime than the reference device with ITO.

Nakada et al. in Yamagata University (Japan) developed OLED devices with an AgNW layer, by using a stacked anode structure of AgNW and conducting polymer [4, 6]. The device structure is shown in Fig. 11.5. Due to the non-uniform hole injection from nanowires, it is difficult to obtain uniform emission when organic layers are directly deposited on the AgNW layer. On the other hand, by coating a conducting polymer on the AgNW layer, uniform emission was obtained. They successfully fabricated a prototype panel with the emission area of $32 \times 32\,\text{mm}$. Although the conductivity of the conducting polymer is not so high and the emitting uniformity of OLED devices with AgNW is an issue, the combination of AgNW and conducting polymer can be one solution as non-ITO transparent electrode.

Furukawa et al. in Yamagata University (Japan) also investigated flexographic printing technologies of silver nanowire [5]. It should be noticed that the required wet thickness on a substrate is thicker than $10\,\mu\text{m}$ for realizing reasonable thickness of the dried film, since the concentration of silver nanowire solution is as low as about 1%. Since such a thickness is not able to be printed by the usual flexographic printing technology, the solvent of the

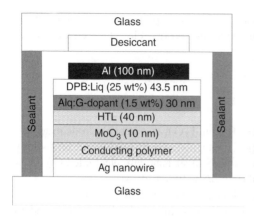

Figure 11.5 Device structure and an emission picture of OLED with AgNW. The device size is 50×50 mm. The emission area is 32×32 mm [4–6]

Figure 11.6 Emission picture of OLED device with the stacked anode consisting of flexography printed AgNW and flexography printed conducting polymer [5]

silver nanowire solution and anilox roll were optimized, and the plate was improved. They fabricated an OLED device, as shown in Fig. 11.6.

11.1.4   Carbon Nanotube (CNT)

Carbon nanotube (CNT) is a promising candidate for non-ITO transparent electrodes because of its unique optical and electric properties, mechanical flexibility, durability, and good adhesion to flexible substrates.

Table 11.3 Typical data of developed transparent conducting films of carbon nano tubes (CNTs)

Affiliation	Material	Transmittance (%)	Resistivity (Ω/sq)	Ref.
Ajou University (South Korea)	SWCNT	80	85	[20]
Aalto University (Finland)	SWCNT	90	110	[21]
Kyushu University	SWCNT	95	120	[22]
Unidym Inc (USA)		91	60	[23]
Rice University	SWCNT	86	471	[24]
SWeNT*	SWCNT	85	400	[25]
AIST	SWCNT	89–98	68–240	[26]
Toray	DWCNT	90	270	[27]

* SWeNT : SouthWest NanoTechnologies
SWCNT = single-wall carbon nanotubes
DWCNT = double-wall carbon nanotubes

Table 11.3 shows some examples of developed transparent conducting CNT films.

Various CNT film fabrication methods have been reported. They are spin-coating [20], vacuum filtration [21, 23], spray coating [22], dip coating [24], doctor-blade method [26], layer-by-layer assembly [28], and spin-spray layer-by-layer assembly [29].

At present, the conductivity of CNT is much higher than normal ITO by about one order. Therefore, it is rather difficult to apply CNT films to OLED devices, while they can be applied to other devices which do not require low resistivity. For example, Oi et al. of Toray Industries (Japan) reported that a twist-ball type electric paper display was fabricated using a flexible film with CNT transparent electrodes [27].

11.2 Organic TFT

Organic TFT may offer a possibility as a future backplane technology, while there are many technical issues which need to be solved before commercialization. Figure 11.7 shows a typical device structure for organic TFT, while various device structures are possible.

In 2004, Chuman et al. of Pioneer (Japan) developed and applied organic TFTs to backplanes for OLEDs. They fabricated a monochrome AM-OLED driven by OTFT with pentacene [31]. Table 11.4 shows the specifications of the OLED panel and OTFT. While the mobility of the transistor was only $0.2\,cm^2/Vs$, luminance of $400\,cd/m^2$ was achieved by using a long channel width.

Organic TFTs can also be used for flexible AM-OLED. One of the advantages of OTFT is low process temperature, which is suitable for many plastic films such as PEN (Polyethylene 2,6-naphthalate). Suzuki et al. of NHK Science & Technology Research Laboratories (Japan) fabricated a 5″ QVGA flexible AM-OLED driven by OTFT, using PEN film [32]. Harrison et al. of Plastic Logic (UK) reported fabricating a 3.86″ flexible bottom emitting AM-OLED display with 100 ppi [33].

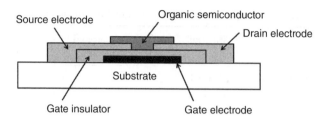

Figure 11.7   Typical device structure of organic TFT [31]

Table 11.4   Specifications and characteristics of AM-OLED panel with organic TFTs [31]

Pixel number		8×8
Pixel pitch		1 mm
TFT design	TFT circuits	$2T + 1C$
	Channel length	$10\,\mu m$
	Channel width	Driving TFT: $680\,\mu m$
		Switching TFT: $400\,\mu m$
	Aperture ratio	27%
TFT characteristics	Mobility	$0.2\,cm^2/Vs$
	Threshold voltage	$-3\,V$
	On/off ratio	1.7×10^4
OLED design		Bottom emission
Luminance		$400\,cd/m^2$

11.3   Wet-Processed TFT

Wet-processed TFT has attractive features because it can contribute to the reduction of process temperature, lower the manufacturing cost, and be an environmentally friendly product. In particular, it is noticed that wet-processed TFT has an excellent compatibility with flexible AM-OLED displays.

One candidate is solution-processed oxide TFTs, which have been widely studied and reported [34–39]. Some of reports are summarized in Table 11.5.

As a process of solution-processed oxide TFT, a solution containing metal oxide based precursors dissolved in organic solvents are spin-coated on a substrate, following by annealing at temperatures such as 350 °C [34] to remove the residual solvents and consolidate the metal oxide film.

Evonik Industries AG developed a solution type metal oxide semiconductor material that shows high mobility of more than $20\,cm^2/Vs$ [34].

Holst Centre/TNO et al. reported a flexible low temperature solution-processed oxide semi-conductor TFT backplane for use in AM-OLED displays [38]. The device was fabricated on a polyimide film deposited on a glass wafer by spin-coating. On this substrate a bottom barrier layer was deposited to prevent penetration of moisture and oxygen. TFT backplanes were fabricated directly on top of the bottom barrier layer. As a semiconductor material, proprietary

Table 11.5   OLED Prototype displays fabricated by wet-processed TFTs

Affiliation	TFT materials	Mobility (cm²/Vs)	Highest process temperate (°C)	Prototype display	Ref.
Holst Centre/TNO, IMEC, ESTA, Evonik	IGZO	2	250	6 cm, QQVGA (85 ppi), monochrome	[38]
AUO*	Metal oxide	4.02	370	4″ QVGA	[39]

* AU Optronics Corporation

iXsenic S metal oxide semiconductor material provided by Evonik Industies AG was spin-coated on the gate insulator and then annealed at 250 °C for one hour under ambient conditions. The semiconductor film was patterned by a standard photolithographic process with oxalic acid as an etching agent. On the anode combined with TFT, top emitting OLED was fabricated with a transparent cathode and then encapsulated by a top barrier layer. The TFT circuit architecture was two transistors and one capacitor. After the encapsulation, the glass substrate was de-bonded. The mobilities of TFT were greater than 2 cm²/Vs. They reported a 6 cm monochrome QQVGA prototype AM-OLED display, with 85 ppi and a pixel size of 300×300 μm.

AU Optronics Corporation also reported a flexible AM-OLED prototype driven by solution-processed metal oxide TFTs. They reported a mobility of 4.02 cm²/Vs [39]. They fabricated a 4″ QVGA full-color flexible AM-OLED display using a solution-based metal oxide semiconductor and organic dielectric layer.

Another candidate for solution TFTs is organic TFTs.

Tokito et al. of Yamagata University (Japan) successfully fabricated a fully printed OTFT array (30×30 dots). The fabrication process is shown in Fig. 11.8 [40]. A PEN film with a thickness of 125 μm was bonded on a glass substrate using adhesive. On the PEN film, cross-linked poly(4-vinylphenol) (C-PVP) was spin-coated as a planarization layer. The gate electrode was printed by ink-jet, using a silver nanoparticle ink (JAGLT-01) provided by DIC. Cross-linked poly(4-vinylphenol) (C-PVP) and poly(p-xylylene) (parylene) were spin-coated as a gate insulator. Source and drain electrodes were printed by ink-jet, using a silver nanoparticle ink (NPS-JL) provided by Harima Chemical. Before fabricating a bank structure, pentafluorobenzenethiol (PFBT) self-assembled monolayer (SAM) treatment was done by an immersion method. The bank structure was dispensed using Teflon (Aldrich). In the bank, organic semiconductor material was dispensed. The organic semiconductor material was lisicon (S1200) provided by Merck or their own new material.

In their fully printed OTFT devices with bottom-contact and bottom-gate architecture, the channel length and width were 20 μm and 1000 μm. The highest process temperature was 140 °C. The fully printed OTFT device using their own new organic semiconductor material showed excellent p-type electrical performance. The maximum mobility was 2 cm²/Vs and the average mobility was 1.2 cm²/Vs. They reported that the sub-threshold swings were small and current on/off ratios were over 10^7. They also reported that the uniformity of the device performance within the panel was excellent. They fabricated several prototypes as shown in Figs 11.9 and 11.10.

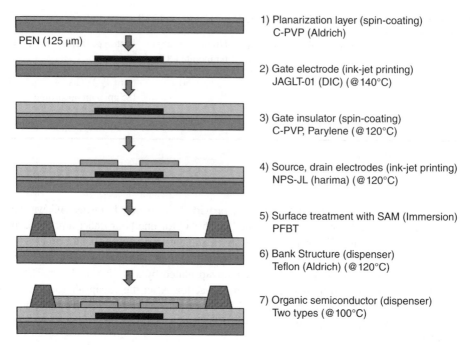

1) Planarization layer (spin-coating)
   C-PVP (Aldrich)

2) Gate electrode (ink-jet printing)
   JAGLT-01 (DIC) (@140°C)

3) Gate insulator (spin-coating)
   C-PVP, Parylene (@120°C)

4) Source, drain electrodes (ink-jet printing)
   NPS-JL (harima) (@120°C)

5) Surface treatment with SAM (Immersion)
   PFBT

6) Bank Structure (dispenser)
   Teflon (Aldrich) (@120°C)

7) Organic semiconductor (dispenser)
   Two types (@100°C)

Figure 11.8   The fabrication process of a fully printed OTFT device on a plastic film substrate [40]

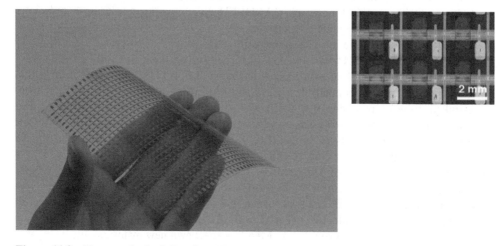

Figure 11.9   Photograph of a fully printed flexible OTFT array (100×100 mm) with 30×30 pixels on PEN film [40]

Kusaka et al. of AIST, Tokyo Electron and DIC (Japan) reported wet-on-wet process for fully printed TFT fabrication [41]. Their TFT is bottom-gate bottom-contact (BGBC) TFTs. All layers were printed using the reverse-offset printing technique. They reported obtaining electron mobility of 3.7×10^{-4} cm^2/Vs, which is only slightly lower than that of the conventional process (7.0×10^{-4} cm^2/Vs).

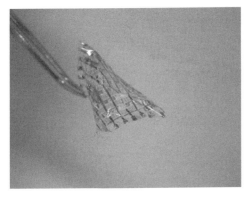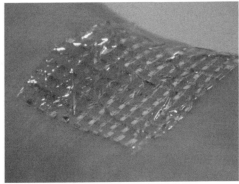

Figure 11.10   Photograph of ultra-thin and flexible OTFT array (10×10) fabricated on a parylene film [40]

11.4   Novel Wet-Processed or Printed OLED

Wet and/or printing processes in OLED fabrication are expected to drastically reduce cost because they do not need expensive vacuum equipment and there is less waste of expensive organic materials. This section reviews some novel technologies of wet-processed or printed OLEDs.

While ink-jet printing technologies have been widely studied for developing OLED displays as described in Section 6.2, alternative printing technologies have also been investigated. They include screen printing [42–44], gravure printing [45, 46], and transfer printing [47]. These printing types are illustrated in Fig. 11.11.

As a printing technology for organic layer, Lee et al. of Sungkyunkwan University (Republic of Korea) reported phosphorescent polymer OLEDs fabricated by screen printing [44]. Their device structure and the emission picture are shown in Fig. 11.12. While the PEDOT:PSS layer was spin-coated, the emission layer was screen-printed. The emission layer consists of a PVK host polymer doped with PBD, α-NPD, and Ir(ppy)$_3$. They used a 400 mesh screen composed of stainless steel fabric of which the diameter was 23 μm and the opening size was 41 μm. The viscosity of the printing ink was less than 2.4 cp for obtaining the layer thickness of 100 nm. The average molecular weight of the PVK polymer was 1,100,000 and the weight of the polymer was strictly controlled to less than 11 mg per 1 ml of chlorobenzene solvent. By using screen printing, they fabricated OLED devices with the maximum efficiency of 63 cd/A.

Gravure printing is a high-throughput printing technique used in the graphics industry. Kopola et al. [45] of VTT (Finland) fabricated a OLED panel with 30 cm^2 by using gravure printing of PEDOT:PSS and blue light emitting polymer.

Chung et al. at Imperial College (UK) investigated gravure contact printing of hole injection and light emitting layers in polymer OLED devices [46].

An alternative method is spray coating. Ishikawa et al. at Kyushu University (Japan) reported a layer-by-layer structure composed by spray deposition [48]. Seike et al. of Kyushu University, etc. (Japan) reported an electrospray method [49].

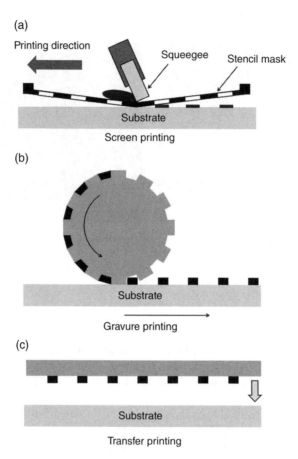

(a)

Printing direction

Squeegee   Stencil mask

Substrate

Screen printing

(b)

Substrate

Gravure printing

(c)

Substrate

Transfer printing

Figure 11.11 Schematic illustration of screen printing, gravure printing, and transfer printing

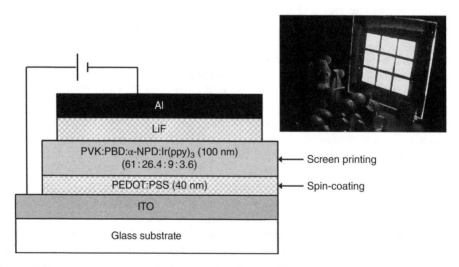

Al

LiF

PVK:PBD:α-NPD:Ir(ppy)$_3$ (100 nm)
(61 : 26.4 : 9 : 3.6)   ← Screen printing

PEDOT:PSS (40 nm)   ← Spin-coating

ITO

Glass substrate

Figure 11.12 Device structure and an emission picture of phosphorescent polymer OLEDs fabricated by screen printing [44]

11.5 Roll-to-Roll Equipment Technologies

The major process in OLEDs and LCDs is currently sheet-to-sheet processes, but from the fabrication cost point of view, roll-to-roll (R2R) processes are attractive. This section describes some examples of R2R equipment technologies which are useful for fabrication of flexible OLEDs.

Tamagaki et al. of Kobe Steel (Japan) reported developing roll-to-roll equipment with sputtering and/or PE-CVD [50]. A picture of the equipment with sputtering and PE-CVD is shown in Fig. 11.13. By using a DC magnetron sputtering method in such equipment, they deposited an ITO film on flexible ultra-thin glass of thickness of 50 μm, width 200 mm, and the length of 10 m. They achieved a good sheet resistance of 7.5 Ω/sq with a thickness of 190 nm. In addition, by using the PE-CVD method in this equipment, they deposited a SiO_x film on a PET film. They also achieved a good WVTR (water vapor transmission rate) of 5×10^{-4} g/m²/day, when the thickness of SiO_x was 500 nm. And they also achieved excellent thickness uniformity on a PET substrate with a width of 1.3 m.

Kim et al. of LG Display (Republic of Korea) reported a roll-to-roll (R2R) deposition-processed amorphous IGZO TFTs [51].

Heya et al. reported roll-to-roll Cat-CVD equipment [52].

Figure 11.13 A roll-to-roll sputtering equipment developed by Kobe Steel [50]

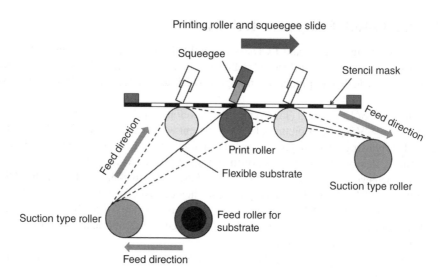

Figure 11.14 Schematic view of the roll-to-roll screen printing equipment with no gap [54]

Hast et al. of VTT Technical Research Centre of Finland reported fabricating OLED devices by using a roll-to-roll gravure printing technology [53].

Kobayashi et al. of Seria (Japan) developed a unique roll-to-roll screen printing equipment for flexible substrates [54]. The unique feature of the equipment was no gap between the screen stencil and the substrate, while previous screen printing equipment required a gap of around 1–2 mm. Figure 11.14 shows the schematic view of the developed screen printing equipment. Due to there being no gap, accuracy and stability of the printing position were drastically improved.

11.6 Quantum Dot

Quantum dot (QD) is another promising material having unique features because they have narrow band emission, high photoluminescent efficiency, easy color control, and solution processability. This section describes the current status of QD materials and their application to light emitting devices.

Quantum dots (QDs) are nanometer size semiconductor particles, in which the band gap is correlated to the size, shape, and composition of the particles. Figure 11.15 shows a schematic illustration of the relationship between the band gap and the particle size. The band gap decreases with increasing particle size. This is called the "quantum size effect". Since the band gap is correlated to emission wavelength, the emission color can be controlled by the particle size [55]. Therefore, by preparing QDs with small fluctuations of particle size, pure color emission with sharp emission spectra can be obtained. Typical particle sizes of quantum dots giving visible emissions are 2–8 nm. For example, a 3 nm QD emits saturated green light (λ_{max} of ~535 nm FWHM of ~30 nm) and a 7 nm QD emits saturated red light (λ_{max} of ~630 nm FWHM of ~35 nm) [56]. In a sense, QDs are highly efficient phosphor crystals.

While there are several types of QDs, a typical material is core-shell type QDs such as CdSe/ZnS. In addition, for wet process compatibility, colloidal quantum dots have been

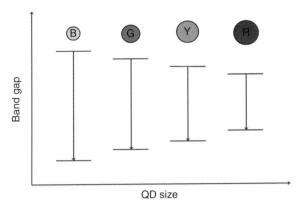

Figure 11.15 Schematic illustration of the relationship between the band gap and the particle size in quantum dots (QDs)

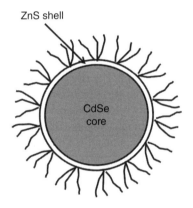

Figure 11.16 Core/shell structure of quantum dots

developed. Colloidal quantum dots usually have substituents combined on the shell surface as shown in Fig. 11.16.

Due to these features of QDs, they have been studied in their application to displays and lighting.

In their first generation, QDs have been down-conversion materials. For example, quantum dots absorb relatively short wavelength light and emit a narrow spectral distribution with a peak wavelength at longer wavelength, based on the size of the quantum dots. Therefore, by pumping with blue light, green and red QDs emit photons in a narrow spectral distribution with a peak wavelength. By using the down-converting property of QDs, the spectra of back-lights for LCDs become sharper, giving high color purity with a combination of color filter of LCDs [56, 57]. In other word, QDs can make sharp the spectra of LEDs that are used as backlights.

In the second generation, QDs have been applied to light emitting devices in which the device structure is similar to OLEDs but the emission occurs from QDs instead of organic

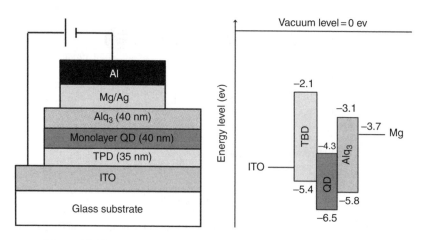

Figure 11.17   The device structure and energy diagram of a QLED [58]

molecules. The device is called a quantum dot light emitting diode (QLED). In addition, due to the wet processability of QDs, QLEDs can be applied to wet and/or printable fabrication process.

Coe et al. of Massachusetts Institute of Technology (USA) reported fabricating QLED devices with colloidal quantum dots [58]. The device structure is shown in Fig. 11.17. The single monolayer of QDs are sandwiched between two organic thin layers. Their QD materials consist of CdSe core, ZnS shell, trioctylphosphine oxide (TOPO) passivation, and trioctylphosphine (TOP) caps. The QD monolayer and the hole-transporting TPD layer were coated by spin-coating. The energy diagram of the QLED is also illustrated in Fig. 11.17. They achieved a 25-fold improvement in luminance efficiency (1.6 cd/A at 2000 cd/m²) over the best previous QLED results.

Kazlas et al. of QD Vision (USA) reported fabricating QLED devices exhibiting peak luminance efficiency exceeding 50 cd/A, luminous power efficiencies greater than 20 lm/W, and operational lifetimes exceeding 300 hours at 1000 cd/m² [59]. They also reported to developing Cd-free QLEDs.

Qasim and coworkers of Southeast University and Nanjing University of Science and Technology (China) reported fabricating a QLED with 10×1 cm area and 32×32 pixels [60].

References

[1] W. H. Kim, A. J. Makinen, N. Nikolov, R. Shashidhar, H. Kim, Z, H, Kafafi, *Appl. Phys. Lett.*, **80(20)**, 3844–3846 (2002).

[2] J. Ouyang, C. -W. Chu, F. -W. Chen, Q. Xu, and Y. Yang, *Adv. Funct. Mater.*, **15(2)**, 203–208 (2005).

[3] K. Fehse, K. Walzer, K. Leo, W. Lövenich, A. Elschner, *Adv. Mater.*, **19**, 441–444 (2007).

[4] M. Koden, H. Kobayashi, T. Moriya, N. Kawamura, T. Furukawa, H. Nakada, *Proc. IDW'14*, FLX6/FLC6–1 (p. 1454) (2014); M. Koden, *The Twenty-second International Workshop on Active-matrix Flatpanel Displays and Devices (AM-FPD 15)*, 2-1 (p. 13) (2015).

[5] T. Furukawa, M. Koden, *Proc. LOPEC (Large-area, Organic & Printed Electronics Convention)*, P3.3 (2015); T. Furukawa, N. Kawamura, J. Inoue, H. Nakada, M. Koden, *SID 2015 Digest*, P-57 (p. 1355) (2015); T. Furukawa, N. Kawamura, M. Sakakibara, M. Koden, *Proc. of IDMC*, S4-4 (2015).

[6] H. Nakada, N. Kawamura, M. Koden, *Proc. of 20th Japanese OLED forum*, S6-3 (2015); T. Yuki, N, Kawamura, H. Nakada, M. Koden, *Proc. of 21th Japanese OLED forum*, S4-9 (2015).

[7] C. Piliego, M. Mazzeo, B. Cortese, R. Cingolani, G. Gigli, *Organic Electronics*, **9**, 401–406 (2008).

[8] J. Ha, J. Park, J. Ha, D. Kim, C. Lee and Y. Hong, *Proc. IDW'13*, OLED4-2 (p. 895) (2013).

[9] S. Ouyang, Y. Xie, Q. Shi, S. Cai, D. Zhu, X. Xu, D. Wang, T. Tan, H. H. Fong, *SID 2014 Digest*, P-147 (p. 1536) (2014)

[10] S. Ouyang, Y. Xie, Q. Shi, S. Cai, D. Zhu, X. Xu, D. Wang, T. Tan, H. H. Fong, J. DeFranco, *SID 2014 Digest*, P-148 (p. 1540) (2014).

[11] K. H. Choi, J. Y. Kim, Y. S. Lee, H. J. Kim, *Thin Solid Films*, **341**, 152–155 (1999).

[12] M. Fahland, P. Karlsson, C. Charton, *Thin Solid Films*, **392**, 334–337 (2001).

[13] E. Bertran, C. Corbella, M. Vives, A. Pinyol, C. Person, I. Porqueras, *Solid State Ionics*, **165**, 139–148 (2003).

[14] C. Guillén, J. Herrero, *Optics Communications*, **282**, 574–578 (2009).

[15] F. Li, Y. Zhang, C. Wu, Z. Lin, B. Zhang, T. Guo, *Vacuum*, **86**, 1895–1897 (2012).

[16] Y. C. Han, M. S. Lim, J. H. Park, K. C. Choi, *Organic Electronics*, **14**, 3437–3443 (2013).

[17] Z. Yu, Q. Zhang, L. Li, Q. Chen, X. Niu, J. Liu, Q. Pei, *Adv. Mater.*, **23**, 664–668 (2011).

[18] J. Jiu, M. Nogi, T Sugahara, T. Tokuno, T. Arai, N. Komoda, K. Suganuma, H. Uchida, K. Shinozaki, *J. Mater. Chem.*, **22**, 23561–23567 (2012).

[19] F. Pschenitzka, *SID 2013 DIGEST*, 61.4 (p. 852) (2013).

[20] J. H. Yim, Y. S. Kim, K. H. Koh and S. Lee, *J. Vac. Sci. Technol. B*, **26**, 851–856 (2008).

[21] A. Kaskela, A. G. Nasibulin, M. Y. Timmermans, B. Aitchison, A. Papadimitratos, Y. Tian, Z. Zhu, H. Jiang, D. P. Brown, A. Zakhidov and E. I. Kauppinen, *Nano Lett.*, **10**, 4349–4355 (2010).

[22] Q. Liu, T. Fujigaya, H.-M. Cheng and N. Nakashima, *J. Am. Chem. Soc.*, **132**, 16581–16586 (2010).

[23] D. S. Hecht, A. M. Heintz, R. Lee, L. Hu, B. Moore, C. Cucksey and S. Risser, *Nanotechnology*, **22**, 075201 (2011).

[24] A. Saha, S. Ghosh, R. B. Weisman and A. A. Marti, *ACS Nano*, **6**, 5727–5734 (2012).

[25] D. J. Arthur, R. P. Silvy, Y. Tan and P. Wallis, *Proc. IDW'13*, FLX3-2 (p. 1534) (2013).

[26] Y. Kim,, Y. Yokota, S. Shimada, R. Azumi, T. Saito, N. Minami, *Proc. IDW'13*, FMC5/FLX1-1 (p. 506) (2013).

[27] T. Oi, H. Nishino, K. Sato, O. Watanabe, S. Honda, M. Suzuki, *Proc. IDW'13*, FLX3-3 (p. 1538) (2013).

[28] S. W. Lee, B.-S. Kim, S. Chen, Y. Shao-Horn and P. T. Hammond, *J. Am. Chem. Soc.*, **131**, 671–679 (2009).

[29] F. S. Gittleson, D. J. Kohn, X. Li and A. D. Taylor, *ACS Nano*, **6**, 3703–3711 (2012).

[30] J. Colegrove, *Information Display*, **30(4)**, 24–27 (2014).

[31] T. Chuman, S. Ohta, S. Miyaguchi, H. Sato, T. Tanabe, Y. Okuda and M. Tsuchida, *SID 04 DIGEST*, 5.1 (p. 45) (2004).

[32] M. Suzuki, H. Fukagawa, G. Motomura, Y. Nakajima, M. Nakata, H. Sato, T. Shimizu, Y. Fujisaki, T. Takei, S. Tokito, T. Yamamoto, H. Fujikake, *Proc. IDW'10*, FLX4/OLED4-4L (p. 1675) (2010).

[33] C. W. M. Harrison, D. K. Garden, I. P. Horne, *SID 2014 Digest*, 20.3L (p. 256) (2014).

[34] S. Botnaraş, D. Weber, D.-V. Pham, J. Steiger and R. Schmechel, *Proc. IDW/AD'12*, AMD5-2 (p. 437) (2012).

[35] M. Rockelé, M. Nag, T. H. Ke, S. Botnaraş, D. Weber, D.-V. Pham, J. Steiger, S. Steudel, K. Myny, S. Schols, B. van der Putten, J. Genoe and P. Heremans, *Proc. IDW/AD'12*, FLX1/AMD2-2 (p. 299) (2012).

[36] J. Steiger, D.-V. Pham, M. Marinkovic, A. Hoppe, A. Neumann, I. Merkulov and R. Anselmann, *Proc. IDW/AD'12*, FLX3-1 (p. 759) (2012).

[37] K.-H. Su, D.-V. Pham, A. Merkulov, A. Hoppe, J. Steiger and R. Anselmann, *Proc. IDW'13*, AMD5-3 (p. 318) (2013).

[38] B. Cobb, F. G. Rodriguez, J. Maas, T. Ellis, J. L. van der Steen, K. Myny, S. Smout, P. Vicca, A. Bhoolokam, M. Rockelé, S. Steudel, P. Heremans, M. Marinkovic, D.-V. Pham, A. Hoppe, J. Steiger, R. Anselmann, G. Gelinck, *SID 2014 Digest*, 13.4 (p. 161) (2014).

[39] L. Y. Lin, C. C. Cheng, C. Y. Liu, M. F. Chiang, P. H. Wu, M. T. Lee, C. Y. Chen, C. C. Chan, C. C. Lin, C. H. Chang, *SID 2014 Digest*, 20.2L (p. 252) (2014).

[40] S. Tokito, Y. Takeda, K. Fukuda and D. Kumaki, *SID 2014 Digest*, 15.1 (p. 180) (2014).

[41] Y. Kusaka, K. Sugihara, M. Koutake, H. Ushijima, *Proc. IDW'13*, FLX2-3 (p. 1526) (p. 1526) (2013).

[42] D. A. Pardo, G. E. Jabbour, N. Peyghambarian, *Adv. Mater.*, **12**, 1249 (2000).

[43] G. E. Jabbour, R. Radspinner, N. Peyghambarian, *IEEE J. Sel. Top. Quantum Electron.*, **7**, 769 (2001).

[44] D.-H. Lee, J. S. Choi, H. Chae, C.-H. Chung, S. M. Cho, *Displays*, **29**, 436–439 (2008).

[45] P. Kopola, M. Tuomikoski, R. Suhonen, A. Maaninen, *Thin Solid Films*, **517**, 5757–5762 (2009).

[46] D.-Y. Chung, J. Huang, D. D. C. Bradley, A. J. Campbell, *Organic Electronics*, **11**, 1088–1095 (2010).

[47] M. Ando, T. Imai, R. Yasumatsu, T. Matsumi, M. Tanaka, T. Hirano, T. Sasaoka, *SID 2012 Digest*, 68.4L (p. 929) (2012).

[48] T. Ishikawa, M. Skakutsui, K. Fujita, T. Tsutsui, *Proc. IDW'03*, OEL3–5 (p. 1321) (2003).

[49] Y. Seike, Y. Koishikawa, M. Kato, K. Miyachi, S. Kurokawa, A. Doi, H. Miyazaki, C. Adachi, *SID 2014 Digest*, P-164L (p. 1593) (2014).

[50] H.Tamagaki, Y. Ikari, N.Ohba, *Surface & Coatings Technology*, **241**, 138–141 (2014); T. Okimoto, Y. Kurokawa, T. Segawa, H. Tamagaki, *Proc. IDW'14*, FLX5-2 (p. 1448) (2014).

[51] K. M. Kim, Y. H. Han, S.-B. Lee, D. Y. Won, Y. H. Kook, S. Choi, C. H. Kim, S. S. Ryu, M.-S. Yang, I.-B. Kang, *SID 2014 Digest*, 13.3L (p. 157) (2014).

[52] Y. Ogawa, K. Ohdaira, T. Oyaidu, H. Matsumura, *Thin Solid Films*, **516**, 611–614 (2008); A. Heya, T. Minamikawa, T. Niki, S. Minami, A. Masuda, H. Umemoto, N. Matsuo, H. Matsumura, *Thin Solid Films*, **516**, 553–557 (2008).

[53] J. Hast, M. Tuomikoski, R. Suhonen, K.-L. Väisänen, M. Välimäki, T. Maaninen, P. Apilo, A. Alastalo, A. Maaninen, *SID 2013 Digest*, 18.1 (p. 192) (2013).

[54] D. Kobayashi, N. Naoi, T. Suzuki, T. Sasaki, T. Furukawa, *Proc. IDW'14*, FLX3-1 (p. 1417) (2014).

[55] M. Bruchez Jr., M. Moronne, P. Gin, S. Weiss, A. P. Alivisatos, *Science*, **281**, 2013–2016 (1998).

[56] J. F. Van Derlofske, J. M. Hillis, A. Lathrop, J. Wheatley, J. Thielen, G. Benoit, *SID 2014 Digest*, 19.1 (p. 237) (2014).

[57] H. Ishino, M. Mike, T. Nakamura, T. Okamura, I. Iwaki, *SID 2014 Digest*, 19.2 (p. 241) (2014).

[58] S. Coe, W.-K. Woo, M. Bawendi, V. Bulovic, *Nature*, **420**, 800–803 (2002).

[59] P. T. Kazlas, Z. Zhou, M. Stevenson, Y. Niu, C. Breen, S.-J. Kim, J. S. Steckel, S. Coe-Sullivan, J. Ritter, *SID 10 Digest*, 32.4 (p. 473) (2010).

[60] K. Qasim, J. Chen, W. Lei, Z. Li, J. Pan, Q. Li, J. Xia, Y. Tu, *SID 2014 Digest*, 7.3 (p. 63) (2014).

Index

Page numbers in **bold** indicate tables; page numbers in *italics* indicate figures.